СОВРЕМЕННОЕ ОБОРУДОВАНИЕ ДЛЯ РАБОТЫ С РАДИОАКТИВНЫМИ ИЗОТОПАМИ

●

Sovremennoe Oborudovanie Dlia Raboty s Radioaktivnymi Isotopami

●

CONTEMPORARY EQUIPMENT
for WORK with
RADIOACTIVE ISOTOPES

CONTEMPORARY EQUIPMENT for WORK with RADIOACTIVE ISOTOPES

Collected Reports

TRANSLATED FROM RUSSIAN

Supplement No .5
of the Soviet Journal of Atomic Energy, 1958

ATOMNAIA ENERGIIA

Press of the State Administration
for the Exploitation of Atomic Energy,
Council of Ministers, USSR, Moscow, 1958

Springer Science+Business Media, LLC 1959

$15.00

Library of Congress Catalog Card Number: 59-14767

Copyright 1959 by Springer Science+Business Media New York
Originally published by Consultants Bureau, Inc in 1959.

ISBN 978-1-4899-4673-7   ISBN 978-1-4899-4671-3 (eBook)
DOI 10.1007/978-1-4899-4671-3

CONTENTS

CONTENTS (continued)

SOME TECHNICAL AND TECHNOLOGICAL ASPECTS
OF THE PRODUCTION OF ISOTOPES
AND LABELED COMPOUNDS IN THE USSR

V. V. Bochkarev, E. E. Kulish and I. F. Tupitsyn

INTRODUCTION

At the present time it is difficult to find a branch of science or technology in which isotopes are not used.

The most valuable of existing methods of producing isotopes is the neutron irradiation of materials in a reactor.

Of the 110 radioactive isotopes to be issued in 1958 in the USSR, 92 are obtained by neutron irradiation.

The present report describes some methods and technological procedures used in the production of isotopes by neutron irradiation and the preparation of labeled compounds from them.

1. Preparation of Materials for Irradiation

The efficient choice of starting materials for irradiation plays an important part in the production of preparations with optimal specific activity and high radiochemical purity, and also in the ease of further processing of the chosen chemical compound.

The first condition is readily satisfied by using the substance in the elementary state for irradiation. However, in a series of cases the fulfillment of the second condition conflicts with the first. Thus, for example, in the production of I^{131} it is more advantageous to use tellurium dioxide, due to the difficulty of dissolving the metal itself, and to neglect the fall in the useful volume of the irradiated material. The same is true in the production of preparations containing P^{32} by the use of elementary (red) phosphorus or phosphorus pentoxide, whose useful volume is only 40%.

Three forms of radiochemical impurity arise during neutron irradiation.

First, radioactive impurities produced by the activation of extraneous elements, which are always present, even in the purest reagents. As a rule, the amount of these radioactive impurities may be reduced to a minimum by preliminary chemical purification of the starting materials. The presence of impurities is detected by means of radioactivation analysis. For example, such a harmful impurity is cobalt in iron-oxide preparations. As simple calculation shows, after irradiation of the iron oxide for 90 days in a neutron flux of 10^{13} neutrons/cm^2 sec, the activity of the cobalt reaches 1% of the total radiation, even when the cobalt content is only 10^{-4}% by weight.

The second source of impurity arises from the simultaneous formation of two or several radioactive isotopes of the same element. For example, let us consider the production of Eu^{155} preparations by the reaction Sm^{154} $(\eta, \gamma) Sm^{155} \xrightarrow{\beta} Eu^{155}$.

The starting material used for this reaction is samarium oxide, which usually contains traces of europium oxide. As calculation shows, the activity of the isotopes Eu^{152} and Eu^{154} formed by the (η, γ) reaction from Eu^{151} and Eu^{153} will not be more than 1% of the activity of the Eu^{155}, if the europium content of the samarium oxide does not exceed 10^{-3}%.

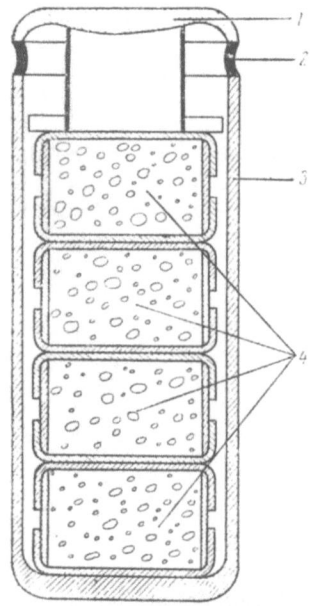

Fig. 1. Cross section of hermetic slug container: 1) lid; 2) place where lid is welded to body; 3) body; 4) small containers with material for irradiation.

The activity of the impurities may be reduced by an efficient choice of the radiation time and the time that the preparation is kept after irradiation. However, this method does not always give the desired results.

Isotopic impurities are sharply decreased when separated stable isotopes are used as the starting materials. Besides giving isotopically pure preparations, the use of isotopically enriched materials makes it possible to increase the specific activity of the preparations.

At the present time it is possible to issue Fe^{55}, Sn^{123}, Te^{127}, Se^{75}, Cd^{115} and other isotopes made from enriched raw materials.

The third and perhaps the most troublesome kind of impurity are isotopes formed as a result of activation of the radioactive isotopes (reactions of the second and higher orders).

Up to 60% of Ta^{183} is formed in Ta^{182} preparations after three months irradiation in a reactor as a result of higher-order reactions. These reactions are particularly strong in the irradiation of isotopes with high-activation cross sections, such as Eu^{152}, for example. Attempts to prepare high-specific-activity Eu^{152} preparations are inevitably unsuccessful due to the "burn-up" of the Eu^{152} isotope with the formation of short-lived europium isotopes.

In the reports of Soviet scientists at the 1st and 2nd International Conferences on the Peaceful Uses of Atomic Energy and in the literature of recent years, there have been detailed descriptions of the constructional features of nuclear reactors operating in the Soviet Union. These reactors are also used for the preparation of radioactive isotopes. Let us only recall that the irradiations may be performed with the materials cooled with normal and heavy water and also in dry channels.

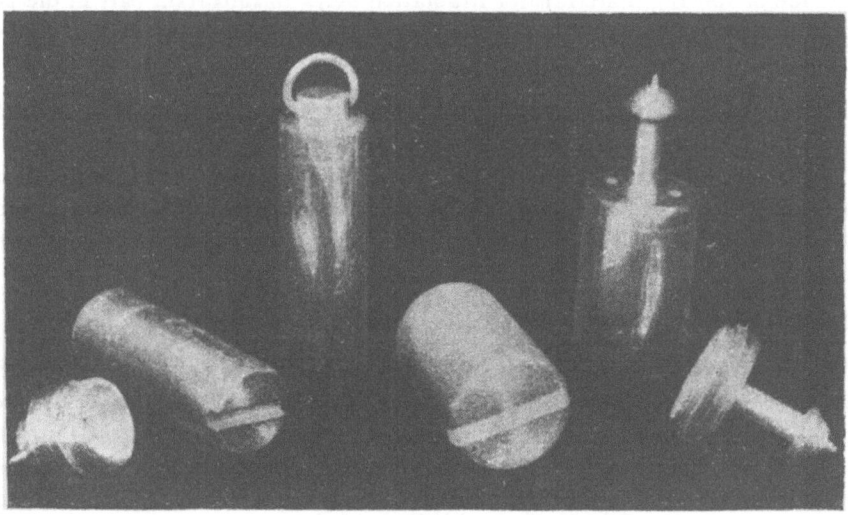

Fig. 2. Nonhermetic slug containers of various types.

To simplify loading into the reactor and unloading after irradiation, all the starting materials are placed in aluminum slug containers. The dimensions of the slug containers are chosen in relation to the constructional features of the reactor and the amount of material to be loaded.

The presence of cooling water in the reactor and the retention and storage of isotopes under water makes it necessary to irradiate the raw material in reliably hermetically sealed casings (Fig. 1). The containers are sealed by means of argon-arc welding. Nonhermetic slug containers with screw-on lids are used for the irradiation of raw material in "dry" channels, in particular in the preparation of short-lived isotopes (Fig. 2).

To reduce the amount of dispensing work in subsequent stages of the processing of irradiated portions, before irradiation, the raw material is divided into standard portions, placed in aluminum containers (volume 0.5, 1.0, 10 and 20 cc) and ampoules of glass which does not contain boron.

Some complications arise in the irradiation of materials with high effective neutron capture cross sections. The high shielding factor of these substances necessitates special measures to ensure uniformity of irradiation. The starting materials are irradiated in the form of thin foils, fixed in the slug containers in a definite way, powdered raw materials distributed in fine ampoules, etc.

2. Irradiation of Samples

The starting materials are irradiated in accordance with the specific requirements of the customers, but the conditions of isotope production in reactors, when they are regularly produced in large amounts, are standardized for maximum results.

Isotopes issued regularly may be divided into the following groups according to the irradiation time of the raw material:

1. Isotopes with a half-life of up to three days (Na^{24}, K^{42}, Ir^{194}). Isotopes of this group, customarily known as "short-lived," are obtained by irradiating materials for 6, 9 and 15 hours.

2. Isotopes with a half-life of from 3 to 30 days (P^{32}, Nd^{147}, Cr^{51}, etc.). The raw materials are irradiated for 30 days.

3. Isotopes with a half-life of more than 30 days (S^{35}, Ca^{45}, etc.). The starting materials are irradiated for 90 days.

4. The isotopes C^{14} and Cl^{36}. The raw materials for the preparation of these isotopes are irradiated for from six months to a year.

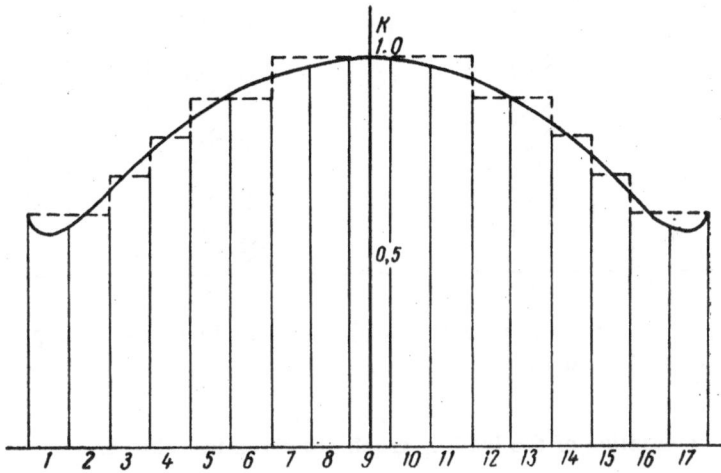

Fig. 3. Distribution of neutron density down the active zone of a reactor (the numbers underneath indicate the number of slug containers successively loaded into the channel).

As is known, the neutron density along a radius and down through a reactor decreases sharply at the edges of the active zone, and, therefore, setting the isotope production channels in positions with a given neutron flux and observing the standard loading graphs are of importance. The standard graphs are usually planned for the greatest mass of isotopes, with reserve positions for the production of rarely needed isotopes. As an example, we will work out the graph for loading the raw material for the preparation of isotopes of the second group (P^{32}, I^{131}, Rb^{86}, etc.) in one of the reactors of the Control Board for the Use of Atomic Energy. The distribution of the neutron density down through the active zone in this reactor is given in Fig. 3. The neutron flux in the center of the channel equals $2.5 \cdot 10^{13}$ neutrons/cm^2·sec. For convenience, we will replace the smooth line of the neutron density distribution by a stepped line, which does not introduce any substantial distortion (the error does not exceed 5-10%).

TABLE 1

Isotope	Saturation activity, A_s in mC/g	Standard activity from tech. spec. A_{st}, in mC/g	Coefficient of position $K = A_{st}/A_s$
P^{32}	900	850	0.95
I^{131}	69	60	0.87
Rb^{86}	650	500	0.77

TABLE 2

Position No. in channel	Coefficient of position K	No. of positions in channel	Isotope
1, 2 16, 17	0.6	4	Other isotopes of 2nd group
3 15	0.7	2	
4 14	0.8	2	Rb^{86}
5, 6 13, 12	0.9	4	I^{131}
7, 8, 9, 10, 11	1.0	5	P^{32}

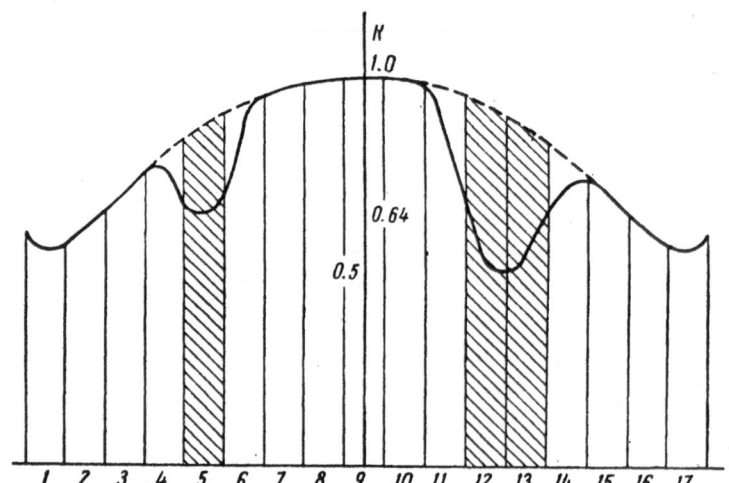

Fig. 4. Fall in neutron density with slug containers loaded in positions 5, 12 and 13.

The initial data for calculation of the loading graph are given in Table 1.

From the data in Table 1 and the neutron density distribution graph, we obtain the standard loading graph for isotopes of the second group (Table 2).

Before the beginning of the irradiation cycle, the working graph for slug-container loading is planned from the standard graph. If the number of positions assigned to any isotope by the standard graph is insufficient, then it may be increased at the expense of isotopes in neighboring positions with higher values of the coefficient K. This does not introduce substantial errors since the flux is not increased by more than 10%.

In the construction of standard graphs, it is necessary to consider the presence of slug containers with strongly absorbing materials, since the latter distort the neutron field (Fig. 4). Due to this, the positions adjacent to these slug containers are left free.

Thus, with an efficient choice of the starting materials, the size of the portions irradiated and the duration and conditions of irradiation, it is possible to standardize the specific and total activity of the isotopes and their compounds issued.

The irradiated materials are either delivered immediately to consumers or sent to laboratories for further processing for the production of labeled compounds and special preparations.

3. Processing of Irradiated Materials. Typical Procedures

The first stage in the processing of irradiated materials from a nuclear reactor or a cyclotron is the radio-chemical isolation of the radioactive isotope. Then the radioactive label is introduced into the molecule of the required chemical compound, which cannot be obtained by direct irradiation. The batches of radioactive preparations obtained are dispensed into standard portions of lower activity.

The following typical procedures are used in the various steps of the production of radioactive preparations:

a) solution of the active materials in water, mineral acids or alkalis with the simultaneous introduction of an oxidant or without it;

b) isolation of a radioactive compound, present in an aqueous solution in macroconcentrations, by reagents lowering the solubility (methyl and ethyl alcohols, acetone, etc.);

c) ion exchange (both for the isolation of isotopes "carrier-free" and for the preparation of labeled inorganic compounds);

d) extraction of radioactive material with water-immiscible solvents;

e) distillation of volatile radioactive substances.

The methods of processing isotopic materials listed may be illustrated by a few examples from the production practice of preparative laboratories.

In addition to the method of preparing Br^{82} described in the literature (extraction from neutron-irradiated organic compounds of bromine, especially, bromobenzene), a technique has been successfully used for the preparation of salts with Br^{82} in which the ion-exchange method has been applied. The use of ion exchange makes it possible to obtain pure neutral solutions of bromides in 85-95% yields from irradiated barium bromide. The process is carried out by remote control in specialized heavy boxes with combined lead and water shielding.

Syntheses of inorganic compounds containing atoms of I^{131} are based on the possibility of distilling elementary iodine (or HI) from aqueous solutions. In particular, the synthesis of potassium iodate, containing the carrier-free isotope I^{131}, begins with the solution of neutron irradiated tellurium dioxide in the minimum amount of 10% KOH solution. To the filtered solution is added excess H_2SO_4, followed by iron sulfate. The elementary iodine is distilled off into a receiver containing 1 mg of KOH in 30 ml of water. The solution of "carrier-free" potassium iodate thus obtained is oxidized with potassium permanganate and the excess of the latter is reduced with alcohol. The manganese dioxide formed is coagulated by evaporation of the colloidal solution to small volume and filtered. The yield of KIO_3 from the isotopic raw material is ~85%. The completeness of oxidation is checked by means of paper chromatography.

4. Procedures for the Preparation of Some "Key" and Complex Organic Compounds

More than 280 of the 450 regularly issued compounds are prepared in the USSR from the isotopes C^{14}, S^{35}, H^3, P^{32} and Cl^{36}. One or two starting substances for each of the isotopes listed are used for the preparation of the labeled compounds, including many of complex structure. The need thus arises for introducing radioactive atoms into determined positions in polyatomic molecules. The conversion to a complex organic product is achieved either by synthesis or by special and radiochemical methods (isotope exchange, "hot" atom reactions and biosynthesis.

The preparation of chemical compounds, labeled with isotopes with soft beta-radiation (C^{14}, S^{35}, H^3, Cl^{36}, etc.), with activities of from hundreds of millicuries to several curies, is performed in laboratories filled with light glove boxes.

The synthetic method has been elaborated to the greatest extent in the production of organic compounds labeled with C^{14}. The starting barium carbonate, labeled with C^{14}, is converted into intermediate "key" compounds (sodium cyanide, barium carbide, methyl alcohol, barium cyanamide, etc.), which, in their turn, serve as starting materials for the preparation of different final products. Without considering in detail the methods of

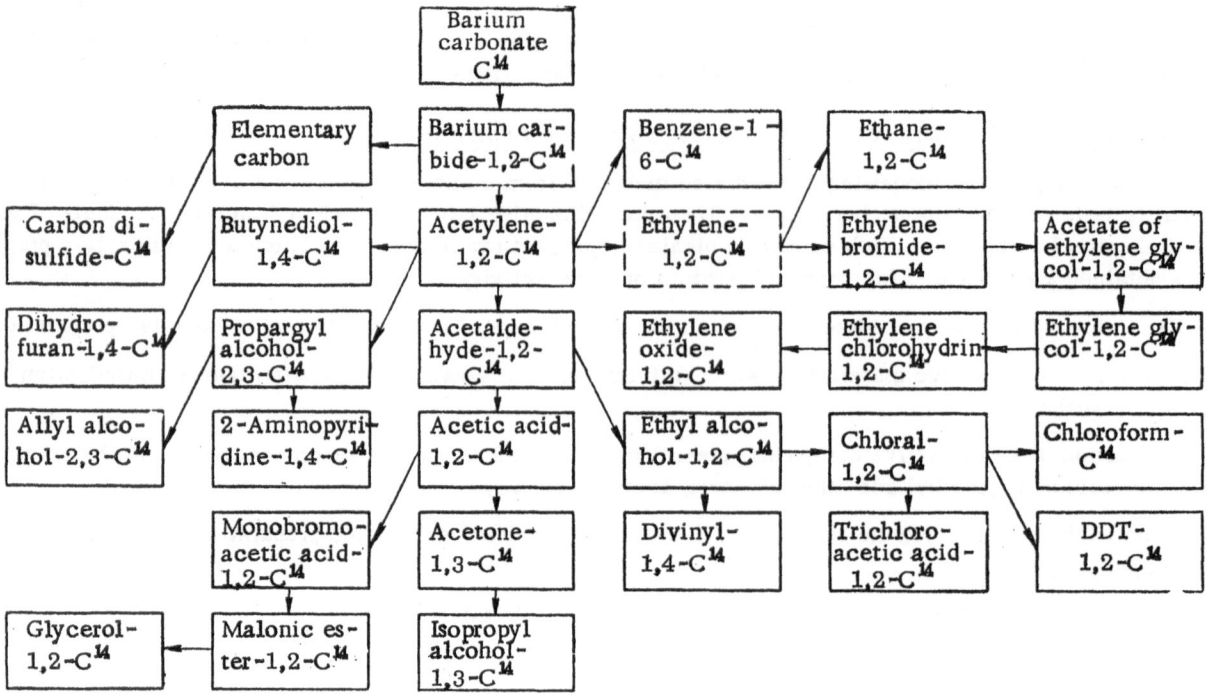

Fig. 5. Scheme of syntheses for compounds based on barium carbide-1,2-C^{14}.

converting the "key" intermediates into the ever-increasing range of organic substances, we will give the synthesis schemes of some important labeled compounds.

Figure 5 shows the branching scheme of syntheses of symmetrically labeled derivatives of acetylene-1,2-C^{14}. As a rule, reactions lying in the framework of the scheme in Fig. 5, are accomplished with efficient use of the isotope, quite simple synthetic procedures and the least expenditure of time, and make it possible to obtain preparations of higher specific activity than other preparative methods. The length of the technological process for the preparation of benzene-1 — 6-C^{14} by polymerization of acetylene is a factor of three or four less than the preparation of benzene-1-C^{14}. In the first case, the yield of product on activity is one and a half times higher. The specific activity of benzene-1 — 6-C^{14} is three to four times higher than the specific activity of benzene-1-C^{14}.

The technology of C^{14}-labeled compound production shown in Fig. 5 provides for the production of from tens of millicuries to several curies of final products in one synthesis. This means that the scale of the syntheses in existing preparative laboratories has been increased by a factor of hundreds in comparison with literature reports. This has made it necessary to design the synthesis processes specially with apparatus requirements in mind.

In increasing the scales of syntheses, it is advantageous to use methods in which reactions involving gaseous labeled compounds are replaced by reactions with solid or liquid intermediates. In this connection we will mention methods of preparing elementary carbon, labeled with C^{14}, and ethylene oxide-1,2-C^{14}. Previously, elementary carbon was prepared in one millimole amounts by the high-temperature reduction of carbon dioxide containing C^{14} with metallic magnesium. Performing this synthesis under production conditions involves procedural difficulties caused by the formation of active side products (carbon monoxide and calcium carbide), the inconvenience of operating huge vacuum apparatus, etc. The method of preparing elementary carbon adopted for production does not have these drawbacks. C^{14} is obtained by heating labeled barium carbide to 700°C in an atmosphere of inactive carbon monoxide, fed continuously into the reaction zone from a gasometer. To remove barium and magnesium oxides, the product obtained is treated successively with hydrochloric acid and hot water. The activity yield on the original barium carbonate is 75% in a good preparation. The isotopic exchange between the elementary carbon-C^{14} and the inactive carbon monoxide during the reaction is insignificant (less than 1%).

A typical synthesis scheme for representatives of various classes of aliphatic compounds, based on monobasic acids-1-C^{14}, is shown in Fig. 6. The methods of synthesizing the bulk of the compounds presented in this

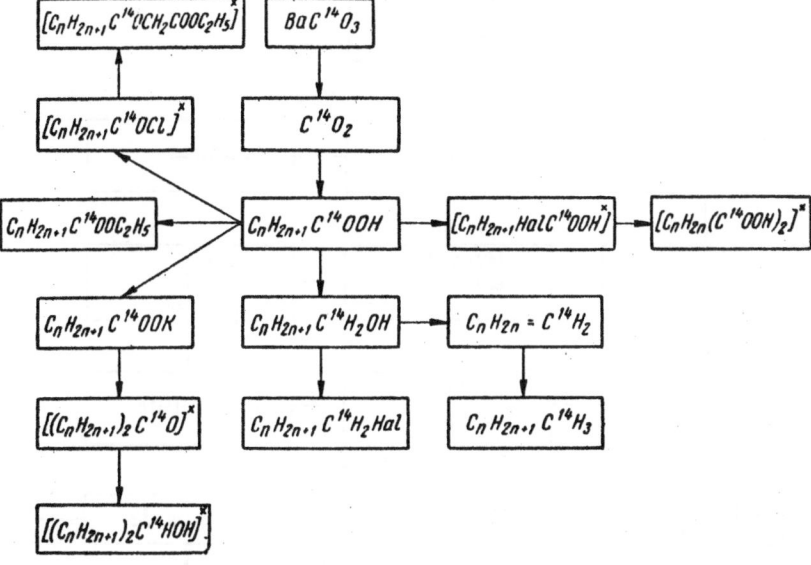

Fig. 6. Scheme of syntheses of organic compounds on the basis of ali-
phatic carboxylic acids-1-C^{14}.

scheme are described in the literature. In the techniques that we have adopted, the scale of the production is increased and separate intermediate steps of the synthesis are omitted. In particular, the preparation of ali-phatic alcohols-1-C^{14} is achieved by a single procedure — direct reduction of the corresponding acid with lithium aluminum hydride (without the usual esterification of the acid). This considerably increases the yields of alco-hols.

The introduction of a C^{14} atom into position 1 of the benzene ring is achieved by condensation of penta-methylene dibromide with ethyl acetate-1-C^{14} by a method, which makes it possible to prepare in successive reactions toluene, benzoic acid and benzene.

On the basis of the simple organic compounds, whose preparation schemes are described above, at the pre-sent time we produce a whole series of more complex C^{14}-labeled substances, required for investigations in chemistry, biology and medicine. They include flotation agents, amino acids, monosaccharides, medicinals, insecto-fungicides, plant growth stimulants and polymers. In all, the preparative laboratories issue more than 180 C^{14}-labeled compounds. Work is proceeding on developing the production of new carbon compounds.

The group of preparations with the isotope S^{35} occupies an important position in the production of labeled compounds. More than 50 such preparations are issued.

The first stage in the processing of radioactive sulfur from irradiated potassium chloride into different sulfur-containing compounds is the conversion of the sulfur into barium sulfate, BaS^{35}O$_4$, and reduction of this with hydrogen to BaS^{35}. Then, oxidation of hydrogen sulfide with potassium ferricyanide gives elementary sulfur in 95-97% yield with a content of the main substance of more than 99% Due to the high specific activity of the isotopic raw material (specific activity of BaS and elementary sulfur of up to 10-20 curies/g), the bulk of the sulfur compounds are synthesized on a small scale; tenths or hundredths of a gram, or in individual cases, grams of the starting materials are used in one operation. For more complete use of the isotope in the synthesis of labeled compounds, obtained by the direct interaction of elementary radio-sulfur with various unlabeled sub-stances, the latter are always used in slight excess.

The preparations are isolated and purified by precipitation (Na$_2$S$_2 \cdot$ 6H$_2$O, Na$_2$S$_2$O$_3 \cdot$ 5H$_2$O), extraction with organic solvents (extraction of KCNS with acetone) and distillation (S$_2$Cl$_2$, P$_2$S$_5$). The synthesis conditions for the preparation of the unlabeled analogs have been modified for the active preparations to give high yields and compound purity and are presented in the scheme in Fig. 7. Thus, the activity yield of potassium thiocyanate-S^{35}, calculated on elementary sulfur, exceeds 95%, that of sodium disulfide Na$_2$S$_2 \cdot$ 6H$_2$O — 90%, sodium thiosul-fate — Na$_2$S$_2$O$_3 \cdot$ 5H$_2$O — 95%, sulfur monochloride (allowing for the recovery of unreacted sulfur) — 95%, etc.

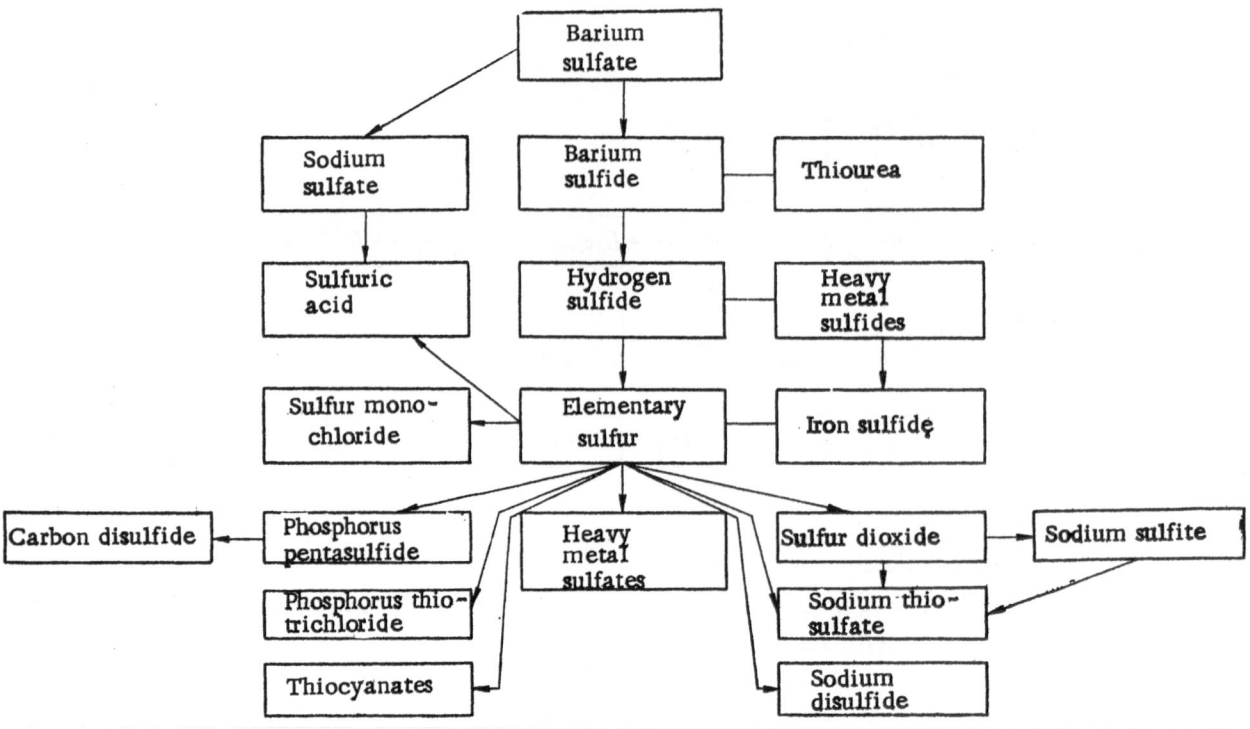

Fig. 7. Scheme of successive syntheses of some S^{35}-labeled compounds.

In their turn, many of the labeled compounds (Fig. 7) serve as starting materials for the synthesis of a wide range of S^{35}-labeled complex compounds, including physiologically active ones.

For example, from thiourea-S^{35} we synthesize methionine, cystein, vitamin B^1, and sulfazole, labeled in the heterocycle; phosphorus thiotrichloride-S^{35} will give thiophos (diethyl-4-nitrophenyl thiophosphate) and metaphos (dimethyl 4-nitrophenyl thiophosphate); sulfuric acid yields sulfamidine preparations (streptocid album and sulfidine); carbon disulfide gives thiuram and alkyl xanthates and sulfur monochloride gives β,β'-dichlorodiethyl sulfide.

In some cases, the radioactive isotope of sulfur S^{35} is introduced into the molecule of the organic compound by isotope exchange between the appropriate substance and elementary radioactive sulfur. In this way we obtain high-activity carbon disulfide and sodium thiopental [the sodium salt of 5,5-ethyl-(1-methyl, butyl)-thiobarbituric acid].

Tritio-organic compounds with the label in a definite position in the molecule are synthesized in appropriate laboratories from T_2O and T_2 with any required activity by the methods developed for the preparation of deutero-compounds. The introduction of tritium atoms into molecules is achieved by reduction of unsaturated organic compounds with gaseous tritium or tritiation with lithium aluminum hydride, hydrolysis of metallo-organic compounds and hydration with tritium oxide. In the case of tritium, the synthesis procedures are simpler than in the case of deuterium, as here it is not necessary to eliminate isotope exchange during the synthesis. Tritiated aromatic hydrocarbons with nonspecific labeling (for example, tritium-labeled benzene and naphthalene) are prepared by isotope exchange of hydrogen between a normal aromatic hydrocarbon and liquid ammonia, labeled with tritium, in the presence of potassamide as catalyst.

5. Shielded Boxes and Some Equipment and Devices for Manipulation

The exhausted boxes and cupboards used in preparative laboratories are equipped with manipulators or gloves, sealed, fitted with aerosol and gas filters and supplied with water, gas, vacuum, electric mains and communications for a drain and the removal of waste.

Usually two types of exhausted, shielded boxes and cupboards are used: a general type with readily demountable, removable apparatus or with general-purpose equipment for carrying out different processes; and

Fig. 8. A specialized heavy box (front view) with combined lead and water shielding for the production of Br^{82} preparations. Water layer thickness — 100 cm: 1) large pneumatic holding device; 2) key for securing the slug container in the clamp of the lifting turntable; 3) lifting turntable; 4) small pneumatic holding device; 5) equipment for opening ampoule; 6) hydromanipulator.

specialized ones, designed for strictly determined technological operations, with stationary apparatus and fittings. Glove boxes with interchangeable apparatus are used in work with carbon, sulfur and other beta-emitters. Specialized boxes are used mainly for the regular issue of a large number of batches of preparations with gamma-emitters and some preparations for medical purposes. Figures 8, 9, 10 and 11 show shielded boxes with stationary equipment for the preparation and dispensing of bromides with Br^{82} and radioactive colloidal gold, for weighing solid beta-preparations and a dispensing box for beta-active solutions.

The different operations and technological processes in the boxes are controlled with various types of mechanical, electromagnetic and hydraulic manipulators and pneumatic holding devices.

Complex master-slave manipulators are used comparatively little in preparative work, mainly in shielded cupboards of the general type. The whole great extension of techniques in Soviet preparative laboratories has been achieved with simple manipulators, designed for a definite range of operations. They can be used for all the possible manipulative purposes with objects measuring from several to hundreds of millimeters and weighing from fractions of a gram to kilograms.

There is also wide use of accessory remote instruments (keys, tweezers, crucible tongs and mirrors on long handles), equipment for remote cutting and opening of slug containers, cans and ampoules with irradiated materials, and also for packing bottles and sealing ampoules with radioactive substances, etc.

Figure 12 shows a photograph of an actual apparatus for remote sealing of ampoules with the controls on the outside wall of the shielded box.

The manipulation of small volumes of active liquids is achieved with hydromanipulators, samplers and automatic burettes and pipettes with remote control. Figure 13 illustrates the principle plan of a hydromanipulator,

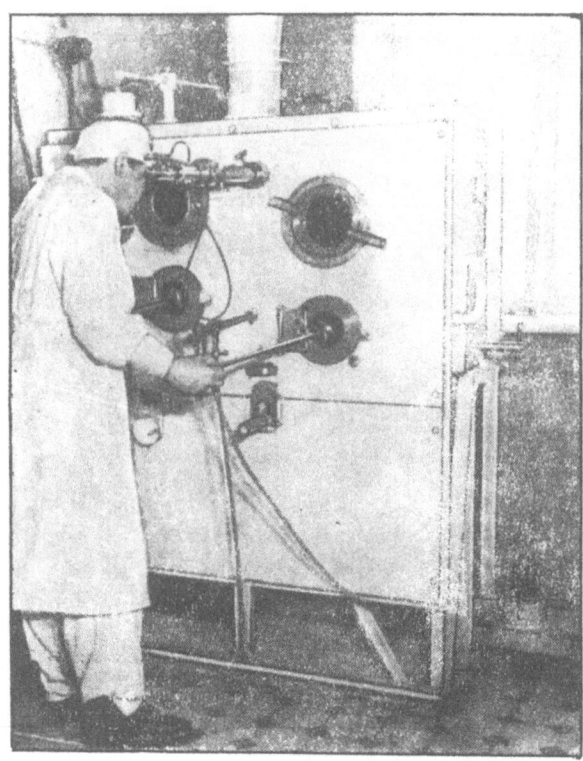

Fig. 9. Shielded box for work with radioactive Au^{198}. From 15-20 curies of Au^{198} is processed in the box at one time.

Fig. 10. Box for weighing solid beta-preparations.

Fig. 11. Box for dispensing beta-active solutions.

which can be used for sampling, transferring accurately measured amounts of solutions from one vessel to another, and dispensing volumes of from 0.1 to 100 ml. Hydromanipulators are fitted with either cable or spring control with the guide tubes in ball-bearings so that the apparatus can be used in a wide zone in the shielded cupboard. Special membrane diaphragms, sensitive to slight pressures, prevent contamination of the hydraulic fluid outside the shield with active solutions.

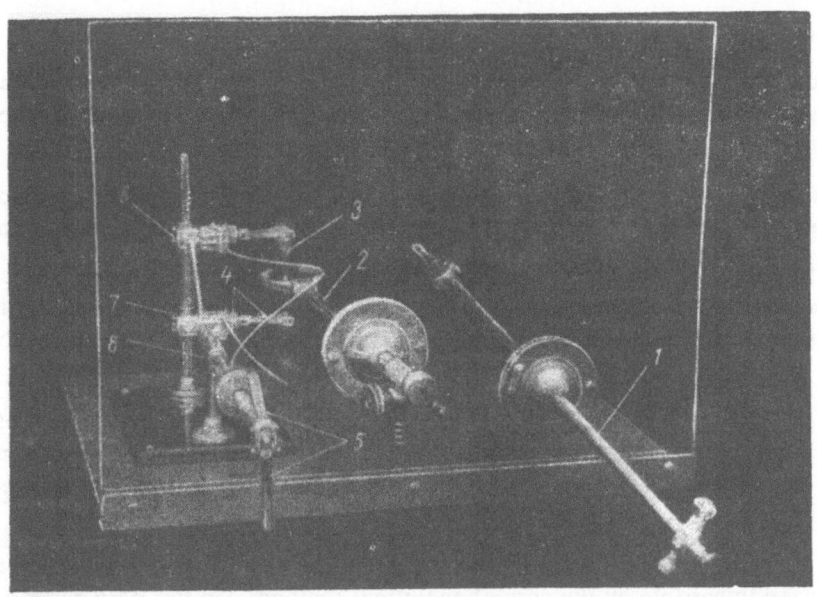

Fig. 12. Apparatus for remote sealing of ampoules: 1) remote holder; 2) gas burner; 3) holder for gripping ampoule; 4) flexible cables; 5) controls for opening clamp; 6) movable holder; 7) fixed holder; 8) horizontal coupling rod.

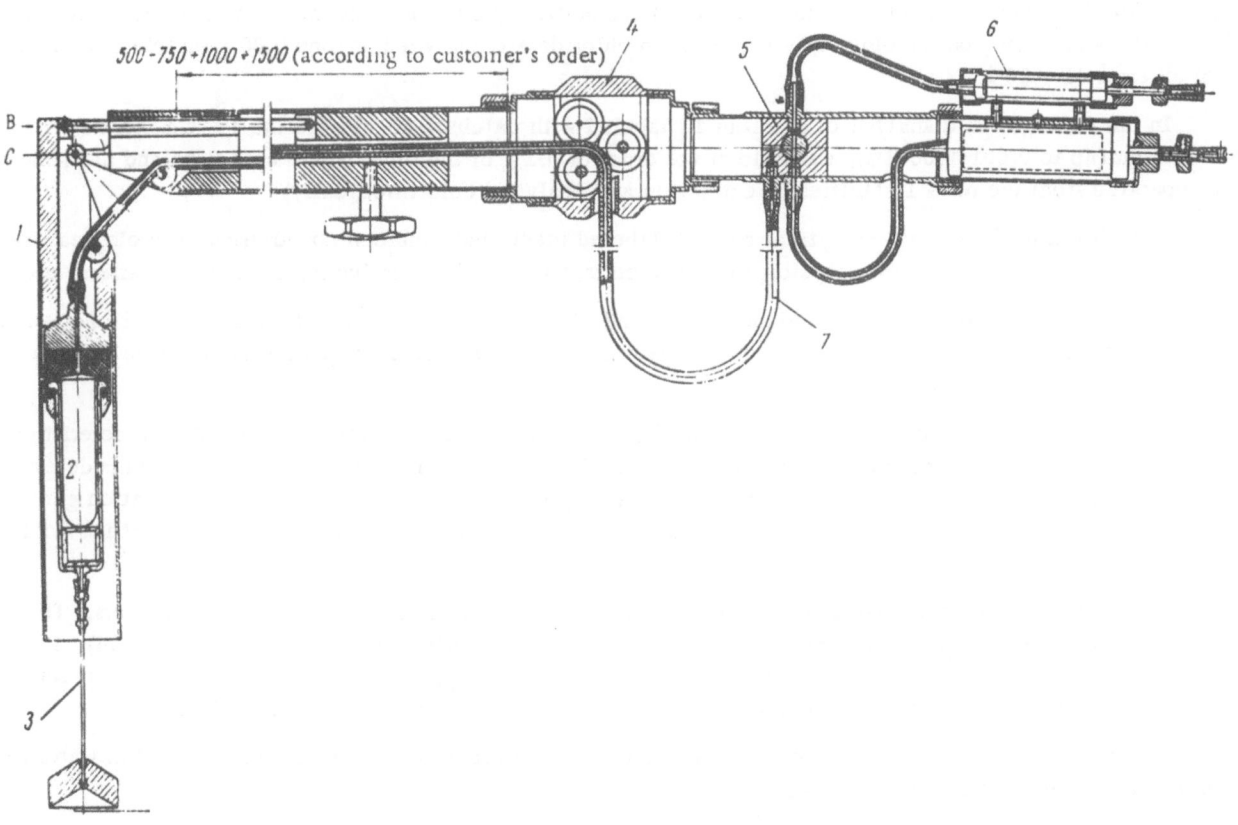

Fig. 13. Principle plan of hydromanipulator.

11

Before dispatch to the consumer, each preparation is checked to find if its basic characteristics correspond to the technical requirements (physicochemical constants, content of the main substance, total and specific activity and content of active and inactive impurities.

The nature of the objects to be analyzed compels the use of purity-control methods for which the amounts of substance required for analyses are minimal and the duration of the analysis is reduced to a minimum.

For the quantitative determination of the basic elements in a radioactive substance, standard, and most frequently of all, volumetric micromethods of analysis are used. Also, the greatest possibilities are offered by those methods in which the chemical analytical control is based on the use of the radioactive properties of the objects analyzed.

We will give two examples in this field. To determine the gamma-isomer content of unpurified hexa-chlorane-Cl^{36}, the sample analyzed is diluted with the chemically pure gamma-isomer of the normal isotopic composition. Then the gamma-isomer is isolated and from the decrease in the original specific activity, we can determine the amount of it in the preparation.

Checking the completeness of oxidation of potassium iodide, containing carrier-free radioactive I^{131}, to potassium iodate proved impossible by the normal analytical methods. In this case paper chromatography was used and the positions of the IO_3 and I zones were detected from the activities of separate portions of the paper strip. A mixture consisting of 80% acetone and 20% 3 N aqueous ammonia solution was used for the chromatography. The zones were first identified by chromatograms of blank, inactive samples.

Spectral and polarographic methods of analysis are particularly effective for the quantitative determination of small concentrations of elements. Comparatively rapid emission analysis is used for determining inadmissible impurities in preparations with short-lived isotopes. To increase the speed of the analyses, certain operations (the introduction of internal standards, etc.) are omitted from the procedure. For example, traces of Ca, Mg, Fe and Pb in pharmacologically pure sodium chloride are analyzed in one 0.05 g sample in an hour with an error of ~10%.

In the polarographic analysis of compounds, labeled with calcium-45, for the Cu, Pb, Cd, As, Fe, W and Mo content (up to 0.001-0.0001%), a sample of the order of 0.1 g of the substance is used (iron and molybdenum are separated from the other impurities by extraction and analyzed colorimetrically).

In addition to the normal analytical control, labeled medicinals and compounds used for biological investigations are tested for possible physiologically important impurities (carcinogenics, arsenic, barium, etc.).

When traces of foreign active substances are found, the preparations are additionally purified. For example, phosphoric acid-P^{32} and its salts are purified from gamma-active impurities by passing the solution through the cation-exchange resin KU-2.

The radiochemical purity of preparations is checked by comparison of their radioactive characteristics (half-life and type and energy of radiation) with tabulated data. The composition of the radiation is determined with beta- and gamma-spectrometers and also by the absorption method. The method of scintillation gamma-spectrometry is of great value in control. It makes it possible to determine rapidly gamma-impurities from one sample of low activity.

The gamma-activity of preparations is measured by comparison with standard gamma-emitters. The measurement of high-activity preparations is performed under water with a special apparatus (Fig. 14). It consists of a chamber, made from plexiglass in the form of a truncated pyramid, at the apex of which is placed the standard emitter and the preparation measured, and in whose base is located an ionization gauge.

Preparations may be placed against the outside of the chamber (sources in ampoules, stored in water reservoirs) or put inside it (normal preparations).

The activity of beta-preparations is measured by absolute counting, using 4π-counters; on special apparatuses with internal filling counters, etc. The activities of preparations are determined in large numbers by a relative method, in which they are compared with standard beta-emitters. The activity of liquid preparations, especially sparingly soluble ones, is also determined on special measuring apparatuses, first calibrated with standard solutions.

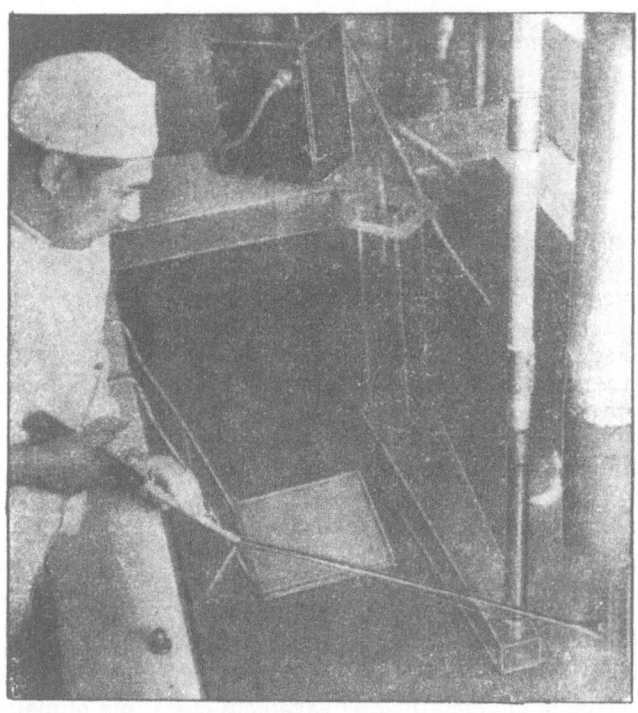

Fig. 14. Apparatus for measuring the activity of gamma-preparations under water.

In all possible cases, the preparation of a sample for measuring the activity of soft beta-emitters is combined with the operations required for analyzing for the main substance in the labeled preparation. For example, the chemical analysis of sulfur-containing preparations (S_2Cl_2, $PSCl_3$, Na_2S_2, ZnS, CoS, metaphos, thiophos, etc.) is performed by quantitative oxidation of all the sulfur and precipitation of it in the form of $BaSO_4$. An aliquot part of the solution, before precipitation of the barium sulfate, is used for measuring the activity of the preparation. The content of the main substance in formaldehyde and acetaldehyde-C^{14} is determined by preparing dimedone precipitates, which are dissolved in alkali for activity measurement.

In conclusion we should note that there is intensive work in the USSR on the further extension of the range of isotopes and labeled compounds issued and the improvement of their quality.

DEVELOPMENT OF REMOTE HANDLING METHODS
IN THE RADIOCHEMICAL LABORATORIES
OF THE ACADEMY OF SCIENCES USSR

G.N. Iakovlev and V.B. Dedov

Over the last fifteen years, as the application of artificial radioactivity has been extended on a broad front, there has been a vigorous development of the techniques and methods of remote handling.

New radiochemical laboratories have been created on the principle of the trizonal system. In these laboratories, the working conditions have been improved considerably, additional measures for dosimetric control introduced and improvements made in the radiationally hazardous operations of maintenance and removal of contaminated equipment and waste. The introduction of the zonal system not only changes the external appearance and the organization of the work of the laboratory, but also presents the problem of a qualitative change in the methods of remote handling themselves. Old mechanical facilities are replaced by modern automatic and telemechanical apparatuses.

In experimental laboratories (where no definite form of product is put out), automation is applied efficiently only to individual, frequently repeated standard operations. Depending on the design, the problem of each radiochemical laboratory is to find those operations to which automatic control can be applied most profitably.

Mechanisms which more or less successfully imitate the action of a man's hand are the basis of remote handling. In the ideal case, the operator should not be conscious of these mechanisms, which are intermediate links between the operator and the object worked with.

Such manipulators may be divided into two main groups. The first group consists of equipment, which to some extent approaches the ideal and is extremely complicated; the mechanical link between the operator and the mechanism which performs the operation, may be replaced here by other types of links. To the second group belongs equipment where all the movements of the mechanism which performs the operation, or at least some of them, do not reproduce the movements of the operator equivalently.

Due to the simplicity of the principles used and the comparatively low cost, the second group of mechanisms is most popular in radiochemical laboratories (Figs. 1 and 2).

Further development of mechanisms of this type is possible through a combination of different movement methods, which offers great possibilities and convenience despite the complication in design. An essential and, in some cases, the main role is played by accessory remote attachments to manipulators and assemblies of devices (Fig. 3).

In some cases, a series of chemical operations is accomplished without the aid of manipulators. This is achieved when only some of the functions of a manipulator are carried out by different forms of remote equipment. The use of such assemblies makes it possible to simplify the construction of manipulators and decrease their amount of movement.

This principle of diversification was developed in the design of automatic chemical apparatuses, where the whole technological process is broken down into a series of elementary operations, accomplished with different assemblies of mechanisms. Electrical manipulators, simple holders attached to arms, usually with not

Fig. 1. Sword-type manipulator in a cylindrical bearing with up to 50° shift of the holder, performing 6 movements, which may be fixed (force is transmitted to the holder with a push rod and a Bowden tube).

Fig. 2. Sword-type manipulator in a ball-bearing with 90° elbow shift of the holder, achieved with a small electric drive (force is transmitted to the holder with metal balls).

more than three movements, transfer the samples along the technological line. Some operations are achieved with manipulators with two movements. All the work in the chemical processing is accomplished with special accessory mechanisms. This sort of setup makes wide use of mobile platforms, tables and discs with collections of instruments and equipment. This type of constructional solution makes it possible to automate the whole technological process easily and this will be discussed below.

In considering the problem of completely hazard-free work, one cannot forget the vital part played by ventilation and the system of waste disposal. These two factors, together with shielding, are the main and in-variable objectives in achieving hazard-free conditions. As practice has shown, inconvenient systems and ill-considered methods of waste disposal can lead to contamination of the working space and unjustifiable irradia-tion of personnel, even when the other services are working well.

The system of laboratory ventilation must fulfill the following three main requirements: 1) the input of conditioned air into the space in accordance with the required change rate and the linear flow rate at the work-ing places; 2) the maintenance of a definite air-movement direction in the building; 3) realization of the re-quired purification and control of the purity of the air ejected into the atmosphere.

Each of the zones is equipped with independent input-output ventilation, with the required pressure dif-ference between the zones maintained with an automatic regulation system with an appropriate emergency shut-off. In order to reduce the degree of contamination of the exhaust air ducts, each of the working positions is fitted with a fiberglass prefilter with a high throughput capacity. Provision is made for periodic washing of these filters at the working position or direct spraying of them during operations.

Fig. 3. Assembly of mechanisms consisting of a transference trolley and chemical equipment attached to independent, movable arms (without the aid of manipulators it is possible to add reagents, decant liquids, remove a sample, mix and remove a decantate): 1) transference trolley; 2) centrifuge; 3) heater; 4) stirrer; 5) vessel for decantation; 6) pipette; 7) dispenser tube.

Fig. 4. Drive for opening doors. A shaft thrust of up to 800 kg is developed. The interlock device shuts off the motor when overloaded. If a breakdown occurs, the drive is easily dismantled from the operating space.

For the removal of liquid waste, two forms of special conduits, made from resistant materials, are used. These conduits are either concreted into the walls or set in special ducts. One line of conduits is used for the removal of highly active solutions and the second for the disposal of contaminated water. The choice of water purification system is determined by the composition of the waste waters, the amount, the required degree of purification and economic considerations. Evaporation and absorption on clays, charcoals and synthetic resins are widely used.

Fig. 5. Well for loading the active sample. Above left — drive of transfer trolley, which is replaced by changing the attachment collar.

An efficient method of disposing of radioactive wastes consists of concentrating them to the maximum and then burying the sludge in concrete blocks.

Approximately 80% of the solid waste consists of combustible materials (paper, plastics, rubber, etc.) and about 5% of glass and metal equipment, which cannot be used further for some reason or other. Processing of the waste consists of reducing the volume to the minimum and compressing it into the form required for burying it for a long period.

In the zonal system, removal of the solid waste and used equipment from the third zone is accomplished by remote withdrawal of the stand on which the equipment is mounted. Such stands are transported along channels in the second zone.

In general, the problems of chemically treating materials are analogous to each other and therefore the designs of radiochemical apparatuses also have some general features.

One of the commonest methods of ensuring reliable operation is the introduction of double lines for this or that process, especially when unique products are being processed.

Doubling is also necessary in the control of apparatus. In case of failure of the automatic system, each apparatus is fitted with a telemechanical control and an obligatory condition of its operation is the presence of a servosystem — the sole means of control of the process occurring.

Electrical functional mechanisms have to be chosen with particular care as they operate under difficult conditions — in chemical chambers and cupboards. There is often a tendency to replace electrical drives by mechanisms based on other operating principles. However, specially sealed electrical mechanisms may be used immediately in the working zone. Electrical drives without special shielding are required by apparatuses outside the active zone.

Besides design features, guaranteeing reliable operation, it is necessary to emphasize the importance of choosing the technological processing scheme itself. Above all, this scheme must be sufficiently reliable and the separate operations readily executed. A study of the various possible technological variations precedes the choice of this or that chemical-processing scheme. Usually, the most promising reagents with the best characteristics are chosen. In particular, such a careful preliminary preparation made it possible to separate weighable amounts of one of the most difficultly separable pairs of elements (americium and curium), when the conditions were particularly drastic and the loading on the resin was 0.01 watt·hour/g.

The main laborious operations in such apparatuses must be automatized. Preparative operations and the final packing of processed products are carried out remotely by an operator, since the problem of setting up an automatic line does not arise. Consequently, the automatic assembly and the compartment for remote operation are the only installations with an internal transference link. This offers great possibilities for accomplishing different technological projects in semihot laboratories of an experimental nature.

Different types of interlock are widely used in apparatuses and, in addition to their particular purpose, these provide means of preventing the penetration of contamination into the operating space. All interlock systems may be concentrated on a special control panel.

Apparatuses with the general characteristics, which have just been considered, are used or planned by the Academy of Sciences of the USSR. They are designed for the isolation of elements by the resolution of mixtures, with the separation and subsequent purification of the required component. These processes represent one of the main problems of radiochemical laboratories and are the first stage in any work concerned with the problems of accumulating radioelements.

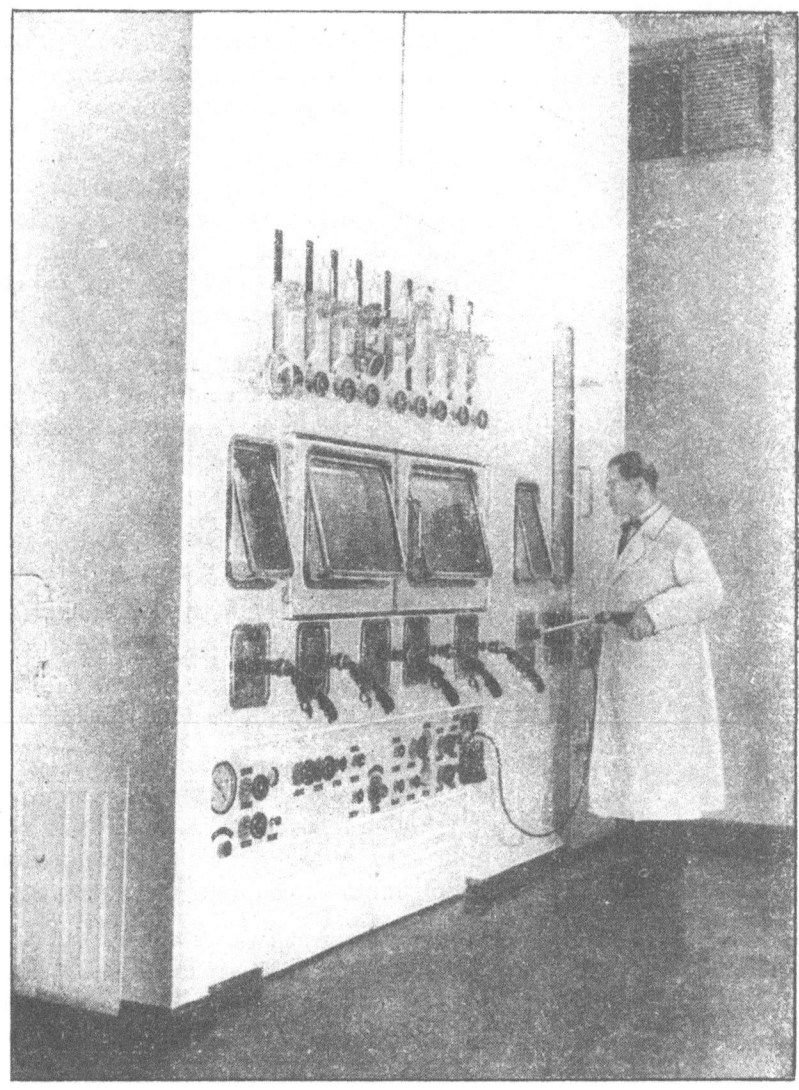

Fig. 6. Automatic chromatographic apparatus (to the left — a general chemical cupboard with remote equipment).

In this report, the main attention will be paid to processes on a laboratory and a pilot plant scale. Industrial technological processes, aimed at producing a definite mass product, will not be considered here.

It is known that three methods are most widely used for separating elements: precipitation, extraction and chromatographic techniques. In the automation of these different types of processes, it is necessary to consider certain general requirements: a) the constructional solution must be sufficiently flexible to allow operations with the maximum range of variations and b) the optimal conditions must be strictly observed during operation.

From this it follows that any automatic device must include elements which maintain the process parameters at a given level. Finding the optimal conditions of the automatic devices themselves is an extremely complicated procedure and is beyond the scope of this review.

Despite the similarity indicated, we have to delimit the processes of element separation. Extraction and chromatographic separation can, to a certain extent, be characterized as continuous and in the simplest cases it is comparatively easy to achieve simultaneous means of stabilizing the physical characteristics (temperature, pressure, flow rate, etc.). In precipitation methods, the technological process is stabilized by a system of programming, which ensures the succession of the operations and fulfills the conditions required for each operation.

Fig. 7. Control panel for automatic chromatographic apparatus with two-channel recording, a relay unit, a radio-electronic circuit and an alarm unit.

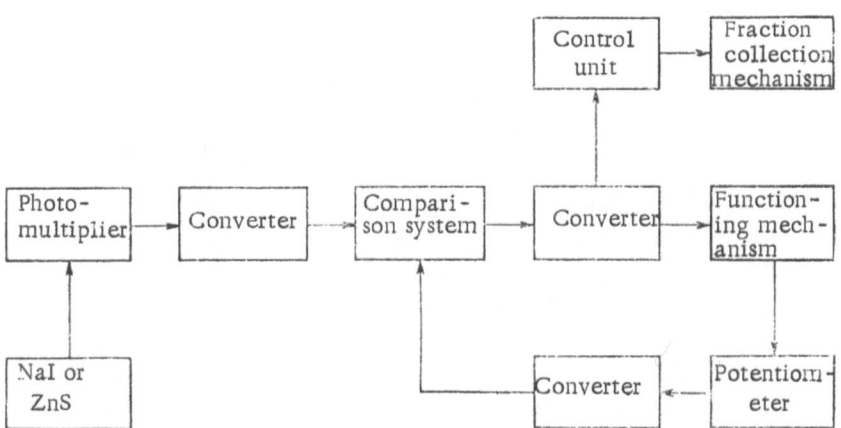

Fig. 8. Simplified automatic system for fraction collection in a chromatographic separation process.

In addition to the characteristics mentioned, in drawing up the program one has to consider such parameters as, for example, the transparency of the solution, the moisture of the precipitate, etc.

Under laboratory conditions, the constancy of the chemical parameters does not usually depend on automatic regulation. It is guaranteed in the preparation of the starting reagents. Here we consider only the approach to the given conditions.

The automatic fittings of extraction and chromatographic separation processes are extremely similar to each other. These processes proceed in the liquid phase and, consequently, it is possible to use one of the methods of scintillation observation, control and registration here.

However, in these cases the possibility of developing closed regulation circuits usually used in automation is difficult, since, as a rule, it is out of the question to construct a feed-back, i.e., a reverse action of the regulating

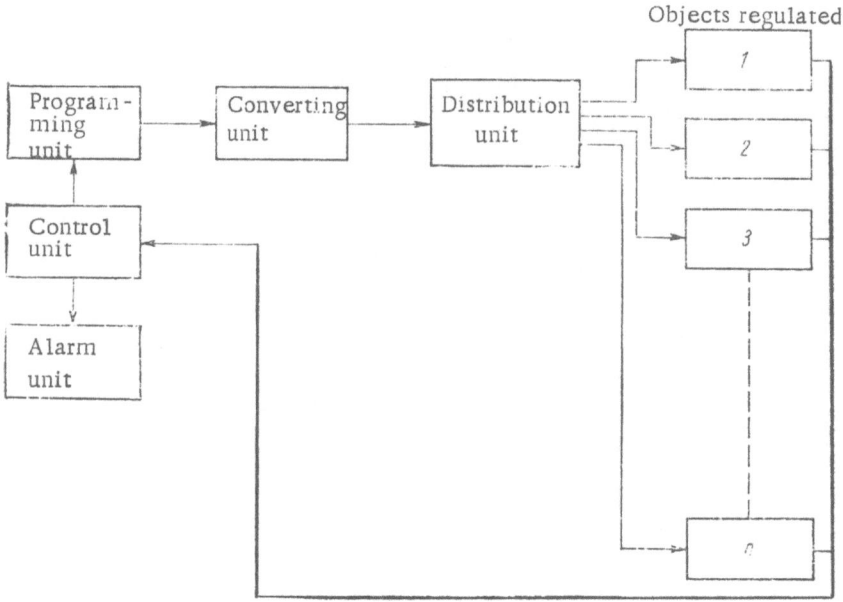

Fig. 9. Simplified automatic system for analytical precipitation operations.

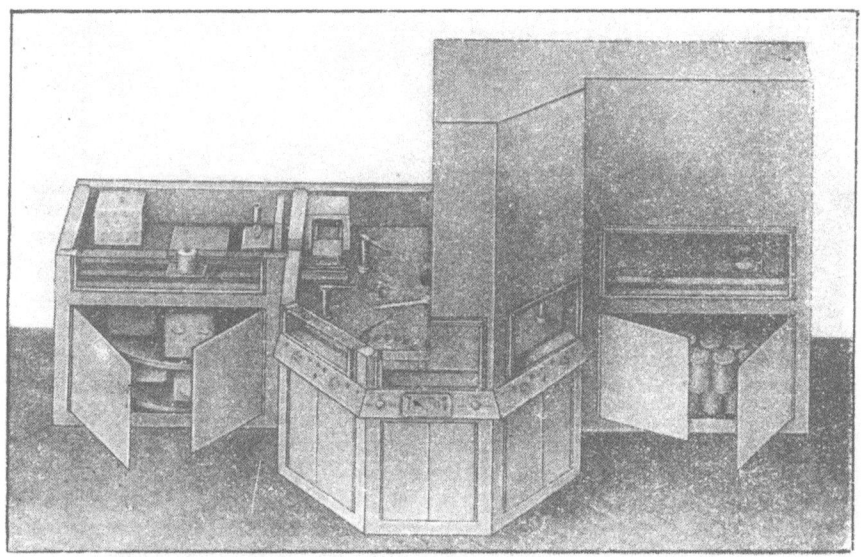

Fig. 10. Sketch plan of apparatus for automatic execution of analytical-chemical precipitation operations.

apparatus on the process itself. This action involves considerable cost. Automation can be applied to operations concerned with the final stages of the processes (collection of the finished separation products) and also to some other functions, which are peculiar to each individual case. When required, the physical parameters can be stabilized automatically.

Since, in a chromatographic process, automatic control can only be applied to physical characteristics, while such important magnitudes as the flow rate at low values are regulated indirectly through the pressure exerted on the eluting liquid, in this case the automation system is very simple.

Figures 4, 5, 6 and 7 illustrate some sections of an automatic chromatographic apparatus.

The system of collecting the finished products is interesting. It is constructed in accordance with the nature of the process, including the sequence of appearance of the elements, whose order is governed by a definite law. In accordance with this, the problems of this system are the collection of the separate elements and the disposal of the residual liquid into a prepared container. This is accomplished with a control unit (Fig. 8). The main function of the comparison system is the determination of the kinks in the curve, characterizing the concentration of the emergent elements. As a result, it is possible to collect fractions of the degree of purity allowed by the chosen separation conditions.

Different principles are used as the basis of automation in available automatic apparatuses for analytical chemical operations by precipitation methods (Figs. 9 and 10). In contrast to the type of automation scheme just described, where the device follows the course of an independently proceeding process and makes appropriate decisions, in this case the control of the process is set beforehand by drawing up a special program. This device is an automatic line with an open circuit. Each successive operation cannot be begun until the previous operation is complete. The fulfillment of these conditions is controlled by special devices.

The program for executing any technological process is set by the operator by means of key switches. The whole technological process is put onto tape as an enciphered code, which makes it possible to turn on the given operations in any number and in any sequence. Programs for technological cycles, whose durations are measured in hours, may be set by qualified operators in a few minutes.

The wide adoption in practical remote work of automatic methods similar to those just described, opens up new possibilities for chemists in all fields, who use not only radioactive, but also stable materials in their work.

SHIELDING AND MANIPULATIVE DEVICES
FOR WORK WITH RADIOACTIVE ISOTOPES

N. V. Samokhvalov

INTRODUCTION

The purpose of this paper is to report on the development and application of new shielding techniques in research and industry.

As it is difficult to describe the details of all existing devices and those being developed within the framework of the proposed paper, the author has limited himself to describing a few of the most important factors in the techniques used for radiopreparative work of a laboratory and production nature. Those interested may find more detailed information on individual problems from special sources.

Naturally, as in every new and fast-developing branch of science and technology, a special terminology has not yet been developed in this field. However, it is undoubtedly needed and should be developed as its lack creates difficulties. In this account there may be certain new and unusual terms, ideas and formulations, but we cannot dispense with them. The specific properties of radioactive materials require special devices to protect the health of personnel working with them.

The preparation, processing and wide application of radioactive isotopes are impossible without the use and further improvement of protective facilities. They should be developed in reasonable conjunction with the existing devices and new ones. One of the most difficult problems, from the technical and practical points of view, is shielding personnel working with high-energy γ-emitters from their harmful effects.

Any technical devices intended for the above purposes must be efficient and convenient in operation, otherwise they will not be used widely. Experience in the development of techniques has shown that the most efficient technical facilities are specialized and not generally applicable devices and instruments. However, specialization must be understood in a wide context, determined by reasonable limits and combined with multifunctionalism of the technical facilities applied to the operation conditions. One of the first indexes of the degree of technical perfection is this multifunctionalism of apparatuses. Specialization of protective facilities does not exclude the possibility of reducing them to a number of standard types, as without this the technical facilities cannot be perfected and utilized widely. Considering present-day practice and the prospects of the near future, it is quite unnecessary to prove the need for perfecting the technical equipment for radiopreparative and radiotherapeutic work under laboratory and production conditions.

Chapter I

DEVELOPMENT OF SHIELDING TECHNIQUES IN RADIOPREPARATIVE OPERATIONS

1. Contemporary State of Techniques in Radiopreparative Operations

The known technical facilities for work with radioactive preparations in laboratory production conditions cover an insufficient range and are often ineffectual and do not satisfy the technical, economic and other requirements.

In many technical laboratory production processes for the treatment and application of radioactive (primarily, γ-active) preparations, adequate shielding facilities have not been adopted.

One of the most widely used types of radiation shielding installations for work with radioactive materials under laboratory production conditions is the exhausted, shielded cupboard.

Such a chamber usually consists of special blocks and sections, which complicate, prolong and raise the cost of planning and constructing the installation itself, as well as the housing, fittings and instruments necessary for normal operation. The use of small lead glass shielding observation blocks in such cupboards greatly hinders visual control of the operation and complicates and raises the cost of constructing such cupboards.

Radiopreparative, radiochemical and radiotherapeutic operations often involve liquid working media, but there are few special remote devices for this purpose and their use is limited.

The various kinds of remote manipulators of the so-called "mechanical hand" type are used, but are not specially designed for operations with liquids, while the specific characteristics and scale of work with liquids under laboratory production conditions require the development and use of special methods and equipment.

The manipulators of the "mechanical hand" type are expensive and complicated to construct and cannot be used for a series of special technical operations involving heavy loads and large spatial transfers or in work in difficultly accessible places and with small volumes of materials. Nevertheless, the advantage of using "mechanical hands" for radiopreparative and radiochemical operations is undeniable.

The combined use in γ-shielded radiopreparative cupboards of simple and more specialized manipulative devices, that are interrelated, considerably increases the range of technical operations which may be carried out under these conditions.

Application of mechanization to, and especially semi- and full automation of, any technical operation and process have always required standardization of the product, techniques, equipment, devices and instruments and also consideration of their interrelation. These problems are being solved in the field of radiopreparative laboratory production operations and radiotherapy techniques.

At present the glass (ampoules and tubes) and light-metal (slug containers and cans) transportation containers and other necessary technical and dispensing vessels for radioactive preparations have not yet been fully standardized.

Standard apparatuses, devices and instruments for standard technical operations, including the dispensing of liquid, powder and granular preparations, sealing and packing of dispensing vessels, loading and unloading of transportation containers, etc., have not been developed sufficiently or have not been accepted.

The problems of technical equipment for typical technical operations, including the primary ones, are especially important as they cause the greatest difficulty to the majority of users of radioactive preparations.

Normally, mechanization of technical processes should in the first instance facilitate the work and increase its productivity.

The conditions of laboratory production work of radiochemists are so unusual that at a certain stage, the mechanization of separate technical operations and even of the processes as a whole sometimes not only does not accelerate, but, on the contrary, complicates and hampers the work.

Most of the known, typical conditions of chemical laboratory production operations have remained essentially the same, but the new property of the chemical preparations, that of radioactivity, has made manual work impossible. This has necessitated essentially new methods and media to ensure safe, remote operation. Execution of the physically light and fine manual work of the laboratory chemist by special remote devices has introduced a series of difficulties.

The working radiochemist needs mechanization not only to prevent harmful after-effects to his own health, but also to that of accessory staff.

At the present time, no one will deny the need for mechanization in radiopreparative and similar work. However, the difficulties facing mechanization, including the development and acceptance of adequate complex methods and devices, result in a negative approach to the role of the new technology in certain cases.

Without acceptance of mechanization this work is intolerable and further progress toward perfecting the technical equipment of the technological processes impossible.

The complicated problem of shielding the human organism from the harmful effects of radioactive materials naturally involves a series of tasks that are themselves of a problematic nature.

The use of radioactive preparations for general medical and therapeutic purposes is in many cases closely related to radiopreparative and radiochemical work.

The relation and specificity of certain operations with radioactive preparations in laboratory production and in clinical conditions (especially with liquid and solid γ-emitters) poses the problem of hand shielding.

The shielding of hands to a great extent ensures the shielding of the whole human organism from the harmful effects of radioactive materials.

This general problem is divided into two categories, depending on the qualitative and quantitative characteristics of the emitters used.

First of all, direct contact of the operator's hands with radioactive materials must be eliminated. This is achieved by health physics measures, overall and specific shielding facilities, special industrial clothing, including various hand and elbow-length gloves, shielded technical vessels for preparations, manual manipulative instrumentation, etc.

Calculations and experiments have shown that such a solution is acceptable for work with mainly α-, β- and γ-emitters with an activity of not more than 10 mg-equiv of radium for a six-hour working day.

It is then necessary to ensure that the hands of the operator are not in an active zone, whose level is above that permissible for a daily norm.

The problem of keeping the hands of the operator out of a zone of radiation above the permissible level is very real and complex. To solve it, first of all special and complicated mechanization of the technical operations, based on modern techniques, must be introduced.

Many of the technical operations with radioactive preparations, especially the primary ones used in research and production, have much in common and this makes it possible to use the same technical measures for them. Thus, for example, chemical processing, dispensing, production of radioactive preparations and other operations must be performed in radiation-shielded preparative cupboards.

Recently, the author and his coworkers have developed and constructed working models and examples of certain appliances for preparative, chemical, dispensing and certain special work under laboratory production conditions and devices for the therapeutic use of radioactive preparations and for research work of a medical character under special conditions. The new equipment was tested under simulated working conditions and was partially adopted in practice.

One of the problems requiring solution was the shielding and fitting of observation windows.

The use of such materials as lead for shielding walls and heavy flint glass (lead glass) for shielded observation blocks in equipment for shielding from high-energy radioactive radiation is inefficient as lead is a material that is unsuitable for construction, expensive and scarce. When it is used in shielding equipment, a strong supporting structure is required, which is difficult to attach to the lead wall blocks and the latter are difficult to connect together.

Shielding from γ-radiation with an energy of the order of 1 Mev and above requires a shielding-layer thickness, which, with other conditions equal, is inversely proportional to the density of the shielding material. The weight of shielding elements (of plane-parallel form) made from different materials replacing lead remains practically the same and the use of shielding equipment of inexpensive and plentiful materials is more advantageous.

The increase in the thickness of the shielding walls in comparison with that of lead is not of vital importance for work in many cases, especially when suitable remote manipulative apparatuses are available (e.g., pneumatic and hydromechanical manipulators, etc.).

Lead glass is even more expensive and scarcer than lead. Its cost in certain types of shielding equipment is up to 50% of the total cost of the equipment. Moreover, observation of the operation through lead-glass observation windows is unsatisfactory as the use of large-sized blocks of this material is practically impossible on a large scale.

In addition, the natural yellowish color of heavy flint glass creates difficulties in following chemical reactions, especially the results of using certain indicators.

A comparison of the cost per unit weight and the shielding properties of certain materials for γ-radiation with an energy of the order of 1 Mev and above gives the following picture. The cost of cast iron is a factor of about 10-20 less than that of lead and its shielding effect is a factor of 1.5-1.75 less. The cost of concrete is approximately one-fifth of that of cast iron and its shielding effect is a factor of 3 less than the latter. Water is several hundred thousand times cheaper than lead glass and has a shielding effect that is a factor of about 5 less. Ordinary technical glass (even the most expensive polished plate glass) is 20-100 times cheaper than lead glass and has a shielding effect which is only a factor of 2 weaker.

High-density liquids, including solutions of heavy metal salts, cost almost the same as heavy flint glass, but have shielding properties between those of ordinary and lead glass.

The use of some "heavy" liquids presents difficulties due to their chemical activity (corrosiveness) and the low stability of the solutions, including inadequate stability toward light and ionizing radiation.

Many operations include the measurement and movement within the space of the radiopreparative cupboard of definite amounts of liquids.

Liquids are measured by volume, using the usual equipment of burets, pipettes, graduated cylinders, measuring glasses, etc., by reading off the volumes on scales within the radiation zone.

In such cases the position, size, thickness, transparency and other characteristics of the observation blocks is of primary importance.

By the use of special methods and manipulative devices, liquid volumes can be measured by scales outside the radiation zone, close to the eyes of the operator, and therefore the thickness and other characteristics of the observation blocks are not as important in many respects.

However, the general working conditions in radiopreparative cupboards require the use of observation blocks of as great a working area as possible, which is particularly hard to achieve when they are made of lead glass.

In adopting the achievements of science and technology for practical use, the economic side of the problem must be considered. What can be used on a relatively small scale for research and experimental purposes is not always feasible for use on an industrial scale.

Considering this, it is necessary to elaborate on the achievements of present-day science and technology and to introduce into laboratory production use standardized large-unit radiation shielding equipment.

Shielding cupboards for γ-emitters should be constructed from large-sized cellular shielding observation blocks with combined shielding of glass, transparent plastic and water.

3. New Devices, Their Purpose and Possibilities

The author has developed a series of interrelated, complementary methods and also pneumatic-hydraulic and electromechanical devices for remote manipulation of radioactive materials in preparative chemical work in shielded, evacuated cupboards. In work under open conditions protection is achieved by distance from the radiation source and the use of screens, particularly, large-sized cellular shielding observation blocks with combined water-glass shielding, which considerably improve the conditions for observation of objects and the lighting inside the chamber.

A reasonable combination of the devices developed results in the minimum movement of the initial containers with the radioactive materials within the given space and makes it possible to keep them in a close-to-vertical position. The advantages of this are apparent, especially in work with liquids.

The improvement in working conditions and visual control broadens the range of technical operations that may be carried out with remote control of the processes.

The main methods and devices, developed under the direction of the author, are listed below.

Fig. 1. Mock-up of γ-shielded preparative cupboard with working equipment (front view): 1) base block; 2) lower front shielding block; 3) table block; 4) cellular shielded front observation block; 5) ceiling block; 6) hydromanipulator with pneumatic control system; 7) pneumatic-mechanical manipulators; 8) manipulator with hydraulic control system; 9) key of self-centering chuck of lifting turntable; 10) special key (articulated socket wrench); 11) mechanical holder; 12) gas torch; 13) control panel for hydraulic column; 14) autotransformer; 15) electrical distribution box; 16) switch for electric motor of lifting turntable; 17) crank of mechanical alternate drive of lifting turntable.

1. A pneumo-hydraulic stationary-manual device for remote manipulations with various radioactive and chemically active liquids in preparative chemical and dispensing operations, carried out in shielded, exhausted cupboards. The use of this device makes it possible to move to any spot within the working space the required liquid doses (even into narrow-necked ampoules at the bottom of containers), to measure them volumetrically by reading a scale which is outside the radiation zone and to count drops.

2. Pneumatic-hydraulic and electromechanical devices for remote measurement, pumping and syphoning of liquids, i.e., moving them only between definite points in the working space.

3. A pneumatic-mechanical device for remote manipulation of radioactive materials, including those in glass and light metal vessels (slug containers and cans), which is used for work in shielded, exhausted cupboards. This device also allows the transfer of the radioactive materials to any spot in the working space, even to those on very different levels.

Fig. 2. Mock-up of γ-shielded preparative cupboard with working
equipment (back view): 1) hydraulic column for remote manipulation of radioactive liquids (on overhead stand); 2) lifting turntable
or stand; 3) light with daylight fluorescent lamp; 4) hydromanipulator; 5) auxiliary electric light fittings; 6) back lower shielding
block; 7) pneumatic holder (with ampoule); 8) transportation container; 9) top of container.

4. Large-sized, cellular, transparent shielding observation blocks with combined glass-liquid shielding
from γ-radiation, whose use is more advantageous than lead-glass shielding.

5. A device for remote dispensing of preparations and the loading and unloading of glass technical, dispensing and transportation vessels, which prevents loss of the contents and contamination of the surroundings during technical and transport operations with liquids.

6. A remote-controlled device for sealing glass ampoules with radioactive materials without waste and
opening them without fragmentation.

7. Principles for dissolving powdered and granular radioactive preparations in the original container rather
than removing, grinding or breaking up the baked and agglomerated radioactive products.

Such operations with powdered and granular materials are difficult to perform by a remotely controlled
technical process as they result in the formation of radioactive aerosols.

Many of these devices were used in a mock-up of a shielded, exhausted cupboard with working equipment
for the production and dispensing of certain radioactive preparations (Figs. 1-3).

The mock-up of the cupboard was designed for technical testing of radiation shielding and manipulative
devices (methods of working with them) and for teaching and demonstrating purposes in the training of operator-
radiochemists and for improving their qualifications.

Fig. 3. Mock-up of γ-shielded preparative cupboard with working equipment: 1) dispensing into glass ampoules using a hydromanipulator with a metal needle tip; 2) breaking the neck of a sealed glass ampoule using a special combined wrench-cutter; 3) unscrewing the top of a slug container (gripped in the self-centering chuck) using the lifting turntable and the articulated socket wrench; 4) transferring powder from an opened glass ampoule into the input vessel of the hydraulic column using a mechanical holder; 5) removing glass ampoule from auxiliary bucket using pneumatic manipulator; 6) removing liquid from a wide-necked vessel (measuring glass) using a hydromanipulator with a glass needle tip; 7) transferring a flask with liquid using a pneumatic manipulator; 8) a glass ampoule with liquid, sealed by a method preventing wastage.

Such cupboards with appliances and instruments may be used for chemical processing of radioactive preparations and for dispensing, measuring, pumping, syphoning, titrating and filtering liquids. The semi-automatic execution of ion-exchange processes under defined conditions is also possible. In addition, these devices may also be used to open light metal containers. They are used for dispensing, sealing and opening narrow-necked glass ampoules without fragmentation and moving technical and transportation vessels with radioactive preparations to any, even difficultly accessible positions in the working space. Such technical operations are often performed under laboratory production conditions.

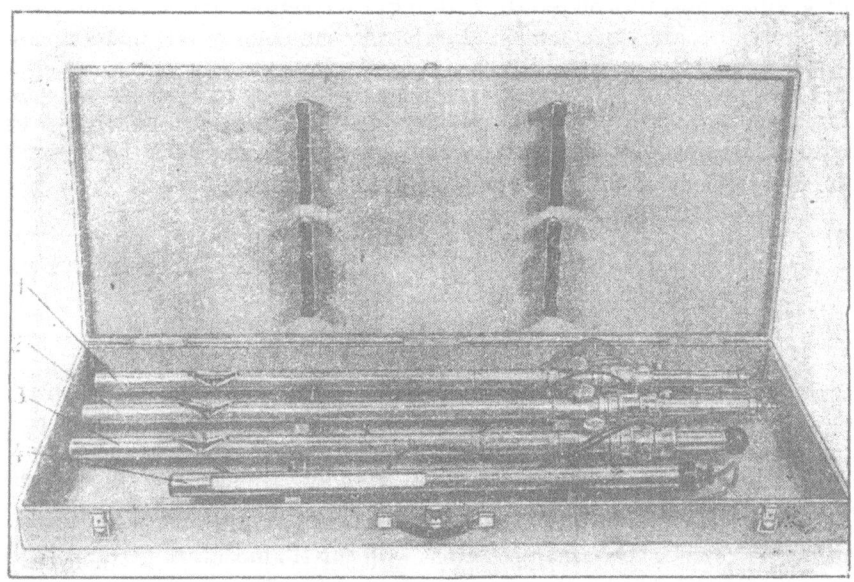

Fig. 4. Pneumatic hydromechanical manipulators and a fluorescent daylight lamp: 1) manipulator with hydraulic control system; 2) hydromanipulator with pneumatic control system; 3) pneumatic mechanical manipulator; 4) fluorescent lamp.

For the remote manipulation of liquids, an experimental model of a stationary manual hydromanipulator was developed and tested under working conditions.

Its use reduces to a minimum and, in some cases, completely eliminates the need for moving through space the initial vessels with liquid work material (Fig. 4).

The following have been developed, tested and partly accepted for use:

1. An experimental model of a semi-stationary hydromanipulator for work with chemically inert radio-active liquids.

2. Experimental models of stationary hydromanipulators with four and eight channel distributors for work with radioactive liquids.

3. Experimental model of a hydraulic column with remote manual and semi-automatic control for measuring, pumping, syphoning and performing ion-exchange processes with liquids under definite conditions and for other special operations (Fig. 2).

4. The first experimental model of a stationary, manual pneumatic-mechanical manipulator for remote work with chemical vessels and light metal (slug containers and cans), glass (ampoules and tubes) and other containers, for extracting and moving radioactive materials directly from the holes of the containers to cupboards and storage and vice versa (Fig. 1).

5. The first experimental model of a stationary electromechanical lifting turntable-stand for preparative chemical and dispensing operations with radioactive materials in shielded, exhausted cupboards; it is intended for opening glass and metal containers, facilitating and accelerating dispensing into ampoules, sealing and opening ampoules and transferring relatively heavy articles (Fig. 2).

6. The first experimental model of a remotely controlled gas torch for sealing glass ampoules with radio-active materials in shielded, exhausted cupboards (Fig. 1).

7. An experimental model of a combined wrench-cutter for remote opening of glass ampoules with radio-active materials (Fig. 3).

8. A movable daylight fluorescent lamp for illuminating the inside of a shielded, exhausted cupboard in

preparative chemical work with radioactive materials, ensuring shadow-free lighting with normal color perception, which is important for visual control of chemical reactions by the color of the product, with the use of many of the widely employed indicators (Figs. 2-4).

9. Instruments have been developed for the remote introduction of doses of liquid γ-active preparations subcutaneously, interstitially, intravenously, etc. into human beings and animals and the injection of nonradioactive preparations, with the recipients in the radiation zone or in a holding chamber.

Chapter II

MECHANICAL HOLDING DEVICES

Hand manipulative transfer-holders are used for semi-remote work with radioactive emitters of certain qualitative and quantitative characteristics.

The holding devices described are used in the chemical and other technological processing of harmful materials under laboratory and production conditions.

Interconnected operations often involve manipulations with small amounts of harmful or harmless (but present in a dangerous zone) solid, liquid or gaseous working materials. The working materials are in normal chemical laboratory vessels of low or average capacity and apparatuses. There are also manipulations with other articles, typical of this sort of work and used in production laboratories.

Chemical laboratory vessels are some of the main and some of the most difficult objects to handle in remote manipulation. The great difference in the geometric forms, weights and positions of the centers of gravity in the empty and full states and the fragility of laboratory vessels, and also psychological factors operating when they are handled with dangerous contents, considerably complicate the work, especially when it is performed with special mechanical appliances of the hand-substitute type.

Available types of laboratory manipulators, intended for remote operations, in particular, manual manipulative holding devices, do not fulfill all the requirements for manipulations with the various articles in laboratory use. Their design, dimensions and principles of gripping and holding laboratory vessels are extremely limited.

The main problem in the production of the manipulative holding devices described is to achieve convenience in remote operations in existing specific production laboratory conditions, including working in shielded, exhausted cupboards through small openings for the hands, in hermetic chambers with comparatively thick elbow-length rubber gloves and also under open conditions. The holding devices must be convenient for work with the necessary assortment of articles in laboratory use under various conditions of use. The number of sizes of holding devices of general and special design must be minimal.

Types of simple manipulative holding devices have been developed for preparative chemical work with α-, β- and γ-emitters on the basis of a classification of types of laboratory vessels and other articles in laboratory use for remote work, theoretical reasons, experiments and generalizations from tests. In work with an assortment of articles of specialized laboratory use, holding devices ensure satisfactory safety, reliability and speed of operation and are simple in manufacture and handling, cheap and readily repaired and freed from specific contaminants. As holding devices have been reduced to a small number of basic types, they are efficient and satisfy the given requirements.

The instruments described may be characterized as manual remote holder-manipulators of the hinged-tongs type with two or four long operating arms, ending in gripping jaws of various forms on one side of the hinge and with handles of a generally accepted form, opened by a spring inside the hinge, on the other.

The manipulators are intended for remote work with laboratory vessels and apparatuses, filled with radioactive, noxious, caustic and poisonous substances, solutions at high temperatures and also for handling other articles in production laboratory use.

Manipulative holding devices protect the hands of the operator from direct contact with the objects handled and thus eliminate the danger of burns and toxic effects.

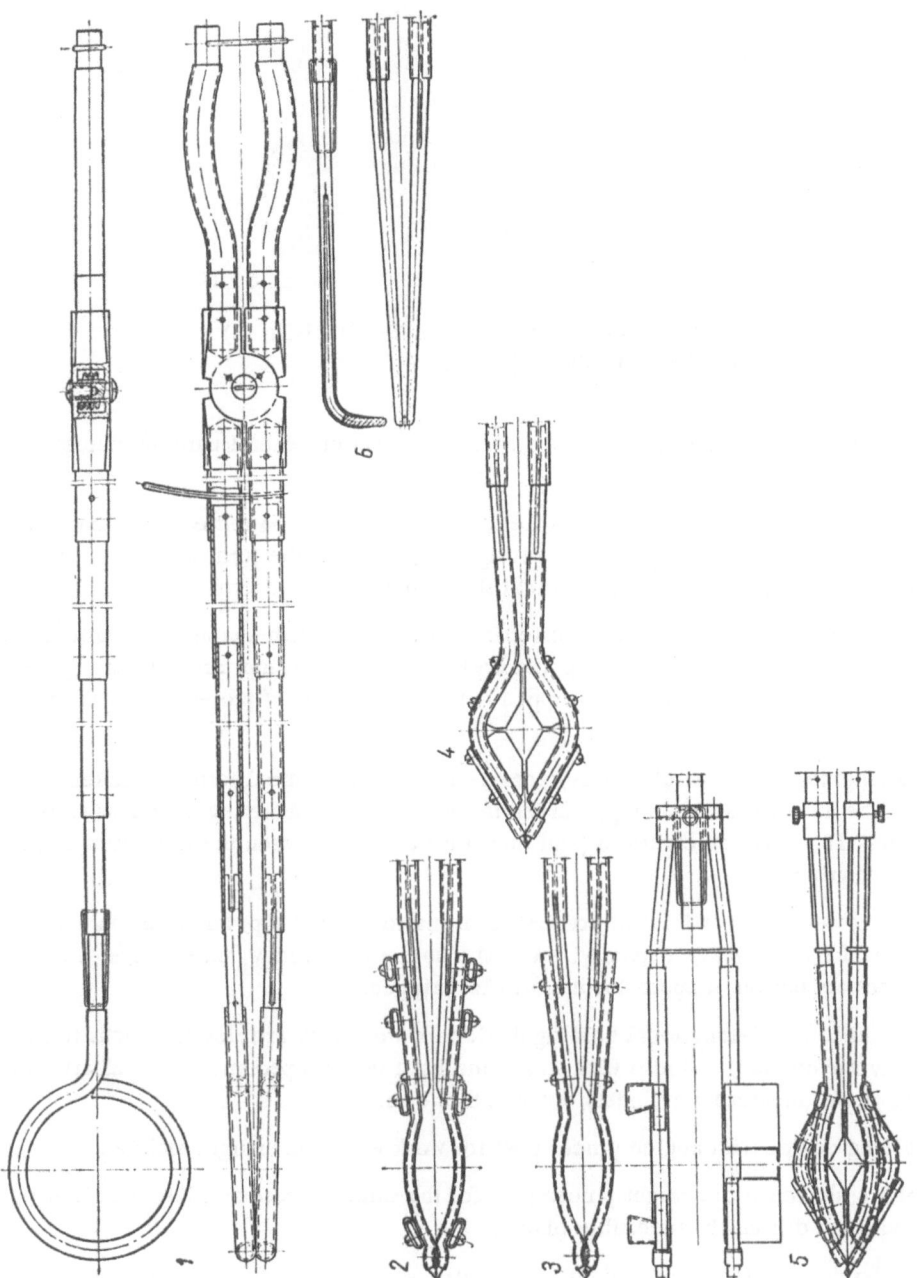

Fig. 5. Types of radiochemical manipulator holding devices.

31

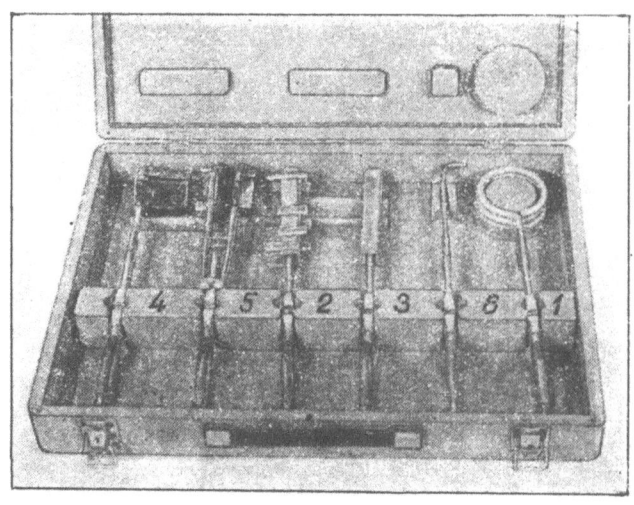

Fig. 6. Set of holding devices: 1) ZPS; 2) ZPCh; 3) ZPTs; 4) ZPK; 5) ZPM; 6) ZPU.

The instruments illustrated in Figs. 5 and 6 form a complete set of general purpose manipulative holding devices.

1. ZPS (Figs. 5 and 6) — a holding device for spherical articles — flat-bottomed and round-bottomed spherical flasks, etc. Main details of the holder: two operating jaws, made in the form of springy split rings, which closely embrace and reliably grip a vessel of a spherical shape and or similar form (Fig. 7).

2. ZPCh — a combined holding device for transferring articles with dish-shaped, cylindrical and inverse conical forms (narrow at the bottom) — evaporating and crystallizing dishes, mortars, beakers, graduated cylinders, etc. The articles to be moved are reliably gripped between two pairs of projecting teeth of different dimensions (Fig. 7).

3. ZPTs — a holding device for articles of cylindrical and inverse conical forms — beakers, test tubes, measuring cylinders, etc. The holder has two springy curved three-stage jaws, differing over a wide range in the operating spans of each stage. This gives an eightfold (and more) difference between the minimum and the maximum diameters of the articles gripped.

4. ZPK — a holding device for articles of conical form (narrow at the top) and also more or less large and heavy conical flasks, beakers of average capacity, etc. The holder has springy, three-stage jaws of an angled form with springy, accessory bottom supports for the articles gripped.

5. ZPM — a combined multi-functional holding device for work with articles of various forms (cylindrical, normal conical and inverse conical) and also with small and light dish-shaped and spherical articles — beakers, conical and spherical flasks, measuring cylinders, evaporating dishes, etc.

6. ZPU — a holding device with angled pincer jaws for work with fine or light articles.

7. ZPP — a holding device with straight pincer jaws for the same purpose as ZPU, but allowing work in deep and narrow spaces and other difficultly accessible places.

The two pincer holders are more convenient for specific working conditions, since they have greater length, a greater jaw span and an efficient operating force.

The length of the ZPM holder may be changed arbitrarily over certain limits by moving the jaws along the main rods. Two pairs of springy gripping jaws, of which the upper pair has an arched form with projecting teeth, will grip the most varied articles. The lower pair of jaws has an angled form with bottom supports (Fig. 8).

With specialized and combined holding devices, it is possible to manipulate two or more articles of the same form and some holders partly overlap in purpose, which reduces the changes during operation to a minimum and gives a reserve for unexpected cases.

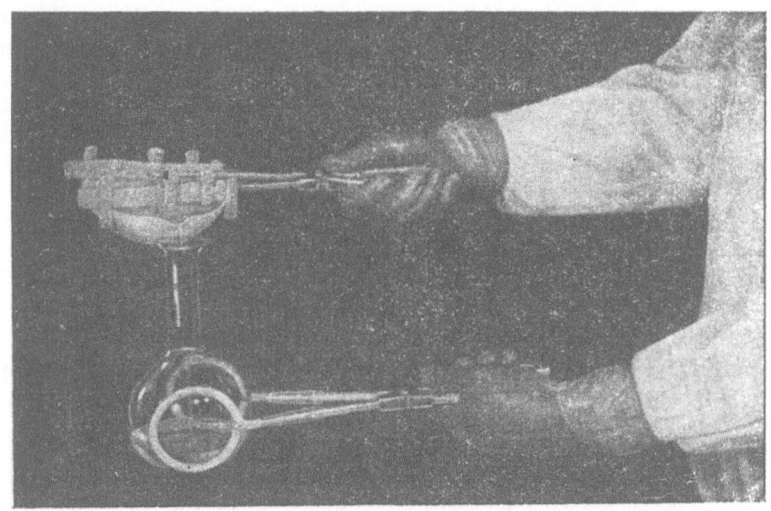

Fig. 7. An example of the simultaneous use of two different holders (ZPCh above and ZPS below). Liquid is poured from an evaporating dish into a spherical flask.

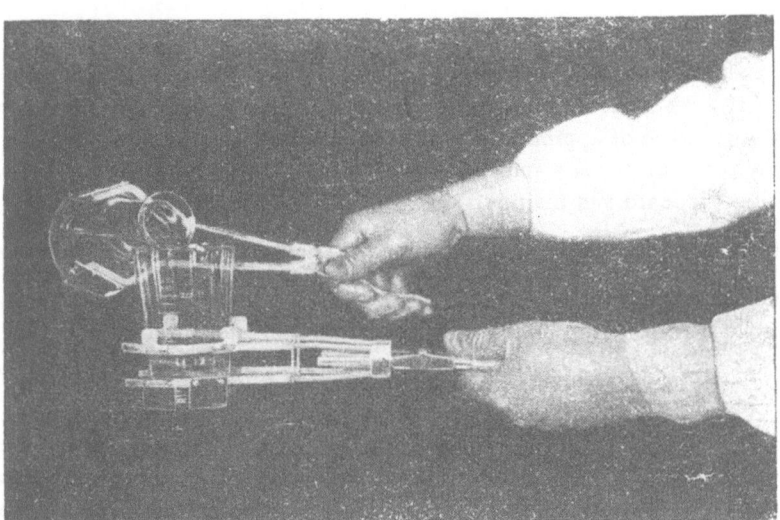

Fig. 8. An example of the simultaneous use of two different holders (ZPK above and ZPM below). Liquid is poured from a conical flask into a measuring cylinder.

The construction of the holders makes it possible to grip articles of any form and in any position in the working space in such a way that the center of gravity of the article remains in the main plane of the holder. This ensures safety and a stable position of the article during work.

Chemical laboratory vessels of low capacity (100 ml and less) and other light articles, handled with a limited number of holders, may be transferred by gripping them by any part.

Manufactured manipulative holders have standard lengths of 350, 500, 700 and 1000 mm, which gives approximately a twofold increase in the protective properties of each successive holder in comparison with the previous one.

All the holding devices are intended for work with objects of different geometric forms, including articles with a flat form over a wide range of dimensions.

When it is necessary to rearrange the holding devices, the operating rods may be interchanged. The holders have a set of spare parts and appliances for decontamination, repair and temporary or permanent increase in the number of the particular type of holder required, etc.

The operating jaws of the holding devices are covered with sections of rubber tubing. Worn and contaminated rubber is readily replaced directly in the working place.

Chapter III

REMOTE PNEUMATIC MANIPULATORS

Remote pneumatic manipulators are intended for small transfers of radioactive preparations in appropriate containers (including fragile, thin-walled ones). One of them is intended for stationary use in shielded cupboards and another, a manual, angled pneumatic holding device, for work of a subsidiary nature, mainly under open conditions.

In the development of these manipulators, the main aim was shielding from γ-radiation, especially high-energy γ-radiations. It was also considered that the most convenient were specialized and not universal technical facilities, in particular, appliances for small transfers of radioactive preparations. The appliances described are part of an assembly of simple technical facilities for remote preparative, dispensing and radiochemical operations.

1. Stationary Pneumatic Manipulator

The stationary pneumatic manipulator is intended for use in γ-shielded preparative and dispensing cupboards for radiochemical work and for the remote execution of a series of typical and special operations, connected with the transfer in space of appropriate vessels with radioactive preparations. In particular, the manipulator may be used for the removal of glass and light metal vessels from containers, transfers of these vessels inside the working space of the cupboard, and manipulations with glass laboratory vessels.

The pneumatic manipulator with its component parts is shown in Fig. 9 in the working position.

The pneumatic holder is the main working part of the pneumatic manipulator and is intended for gripping and transferring different articles to any place within the limits of the working space. It consists of the parts shown in Fig. 10. In the manipulator described, the pneumatic holder is suspended from a flexible connecting hose-lead.

In operation, the pneumatic holder is lowered onto the object to be gripped by it so that the latter is inside the cylindrical bucket. Air is forced through the connecting hose and the bent channel in the bottom of the bucket into the closed space between the walls of the bucket and the sheath. The air presses on the sheath, which squeezes in, encompassing the article tightly and holding it during the subsequent transference of the pneumatic holder in space.

The article is released by discharge of air from the cavity of the pneumatic holder. Atmospheric pressure keeps the sheath flat against the inner walls of the bucket and the article is released.

The article is held in the pneumatic holder by friction between the surfaces of the sheath and the article. Due to its elasticity, under air pressure the rubber sheath encompasses articles of very different forms and dimensions and centers them in the pneumatic holder. The uniform air pressure and the high coefficient of friction of the rubber, even with small operating air pressures, create sufficient force to hold the articles in the pneumatic holder. The pneumatic holder is very simple in design, construction and use.

If the bucket and rubber sheath have sufficiently thin walls, the pneumatic holder may be used for work in the most inaccessible spaces of round and annular cross section. For example, it may be used in the remote loading and unloading of containers with vessels holding radioactive substances.

All this makes the pneumatic holder quite efficient as a special and also a multi-functional gripping device. The pneumatic holder is particularly convenient for work with various fragile articles, e.g., thin-walled laboratory vessels, glass ampoules, etc.

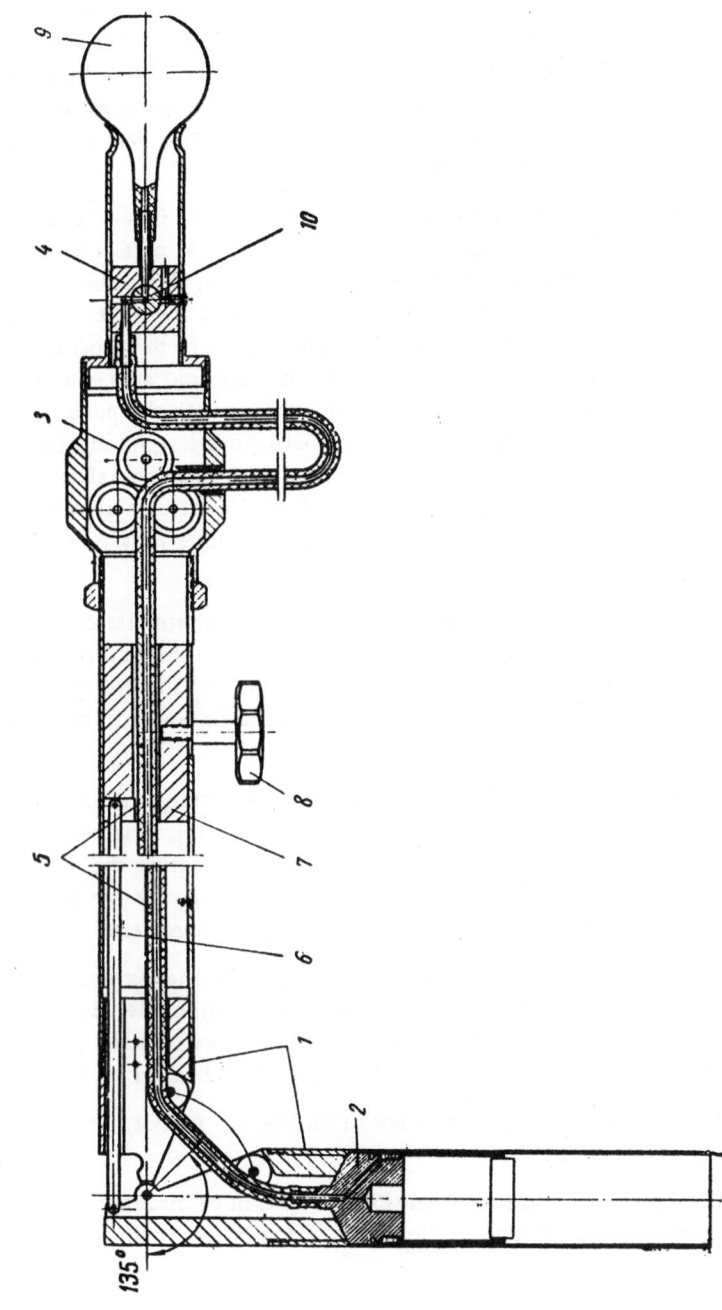

Fig. 9. Pneumatic manipulator for stationary use: 1) hinged tubular mechanism; 2) pneumatic holder; 3) hose extension roller mechanism; 4) pump distributor mechanism; 5) flexible connecting hose; 6) swivel rod; 7) collar; 8) clamping screw of collar; 9) bulb; 10) distributing tap.

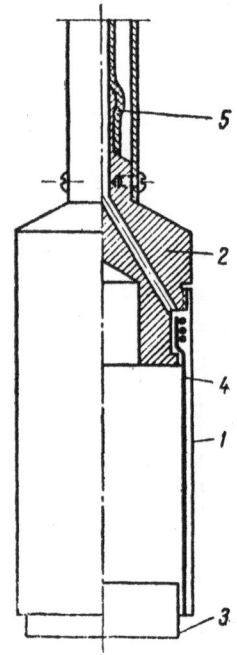

Fig. 10. Single-cavity pneumatic holder: 1) cylindrical bucket; 2) bottom of bucket; 3) sealing ring; 4) cylindrical rubber sheath; 5) pneumatic hose-lead.

The hinged tubular mechanism (Fig. 9) is designed to accommodate (at the ends and inside it) the other parts of the manipulator.

It consists of a round tube with a remotely controlled hinged joint, dividing it into two parts. Such a design makes it possible to give the manipulator a straight or an L-shaped form. The angle between the axes of the two parts may be right, obtuse or acute. Changing the angle of the bend in the main tube results in a change in the operating length of the manipulator, i.e., the distance from the right-hand end to the axis of the pneumatic holder. This is useful during operations. Changing the distance and fixing the angle between the two parts on opposite sides of the hinge are accomplished with a swivel rod, by appropriate shifting and tightening of the collar screw. The collar simultaneously acts as a shielding-sealing plug for the inner channel of the horizontal part of the manipulator tube. If necessary, it may be lengthened.

In using the pneumatic manipulator in shielded, exhausted cupboards, it is desirable that the length of the horizontal part of the tube, from the axis of the hinge to the axis of the clamping screw, should be sufficient to cover the whole depth of the cupboard without changing the right-angular form of the bend in the tube as this length and the right angle are the most convenient for a wide range of operations. It is desirable that the vertical arm of the main manipulator tube should be as long as possible to cover with the pneumatic holder the greatest width of the work space in the cupboard (to the right and left of the manipulator axis) by rotating the manipulator around the horizontal axis. At the same time we should consider the greatest lift of the articles above the working level of the cupboard: the smaller this height, the longer may be the vertical arm of the manipulator and vice versa.

The horizontal arm of the main manipulator tube is best placed in an appropriate guide collar, which holds the tube in any position. Such a collar should be placed at the required height, but not much higher than the head of the operator. The appropriate part of the manipulator is readily inserted into and removed from the guide collar in a straightened position.

The hose extension roller mechanism (Fig. 9) is designed for moving the pneumatic holder in a vertical direction and fixing it, by changing the length of the flexible connecting hose between the pneumatic holder and the rollers of the mechanism. The length of the hose is changed by rotating the crank of the leading rollers of the hose extension mechanism. The pneumatic holder is fixed vertically by the braking of the rollers and the hose due to friction, the elasticity of the hose and the regulation of the rollers.

The mechanism described consists of a hollow cylindrical body and six semi-rollers, arranged in pairs on three parallel axes inside the body (Fig. 9). The gap between the semi-rollers in each pair and the relative position of the roller pairs are regulated with bearing screws. The required relative position of the rollers in relation to the diameter and elasticity of the hose and the required force is achieved by this regulation.

The pump distributor mechanism (Fig. 9) consists of an elastic, rubber, pear-shaped bulb and a special distributing tap, arranged in a cylindrical body.

The mechanism is designed to create an operating pressure in or discharge the air from the pneumatic holder cavity. The air pressure in the pneumatic holder is raised by squeezing the bulb and discharged by releasing the bulb.

In preparing the manipulator for work, some air is removed from its hermetically sealed pneumatic system, consisting of the cavities in the elastic bulb, the pneumatic holder and the pneumatic lead. This is achieved in the following way. The barrel of the distributing tap (Fig. 9) is turned through 90° in a clockwise direction and the bulb squeezed slightly. Part of the air in the pneumatic system is discharged into the atmosphere, after which, the barrel is returned to its previous position. When the bulb is squeezed, the walls of the rubber sheath of the pneumatic holder (Fig. 10) will squeeze into the bucket until they reach the article to be gripped, due to the air pressure inside the pneumatic system. On turning the barrel of the tap through 90° in a counterclockwise direction, the article will be gripped in the pneumatic holder, regardless of whether the bulb is squeezed by the operator or not. The article is freed from the holder by turning the barrel of the tap through 90° in a clockwise direction.

The possibility of holding the article with the pneumatic holder without holding the bulb greatly facilitates subsequent manipulations with the article held. Experiment has shown that with good sealing of the pneumatic system of the manipulator and grinding and greasing of the bearing surfaces of the distributing tap, the pressure created in the pneumatic system may be maintained readily for several days, while the time that the article is held by the pneumatic holder in practice is measured in seconds or minutes.

The hose of the pneumatic manipulator may be made of rubber, rubberized cloth or a suitable plastic. The hose must be quite thick walled (to eliminate kinking and excess extension during operation); it is desirable that the bore of the hose should be not less than 2 mm. The total length of the hose is determined by the required difference in the upper and lower positions of the pneumatic holder and the length of the main tube of the manipulator and may be of the order of several meters. This is necessary in cases where the holder is used to remove radioactive preparations from containers brought under the shielded, exhausted cupboard, in loading deep-storage containers, etc.

All the parts of the pneumatic manipulator may be made of metal. In certain cases, anti-corrosive coatings and finishes may be used on the parts and sometimes the parts are made entirely of stainless steel of an appropriate grade.

2. Manual Angled Pneumatic Holding Device

The manual angled pneumatic holding device is designed for work of a subsidiary nature under open conditions. The operator is protected from harmful effects of radioactive radiations by being at a distance from the source and the short duration of the operation, as well as by using simultaneously accessory shielding media such as stationary and mobile shields, containers, etc.

The manual angled pneumatic holding device actually consists of a pneumatic holder (Fig. 10), a rigid L-shaped tube with arms of different lengths, an elastic, pear-shaped bulb and a connecting hose, located inside the L-shaped tube.

The pneumatic holder is fixed to the short arm of the L-shaped tube by a screw. The elastic bulb is placed with its narrow end in the open end of the long arm of the L-shaped tube, in the same way as it is placed in the body of the pump-distributing mechanism (Fig. 9). The bulb is connected to the pneumatic holder by a rubber hose and a metal connecting tube, firmly inserted into the hose and bulb. The length of the rubber connecting hose is slightly less than the total length of the two arms of the L-shaped tube. During operation, the hose is extended and as a result it holds the bulb in position. This method of holding the bulb leaves it free.

The length of the long arm of the main L-shaped tube is chosen in accordance with the actual working conditions. The most convenient tubes are 3-5 meters long. In order to decrease the weight without decreasing the strength and reducing the length, in certain cases the tubes are designed and constructed as a cantilever of equal bending strength, from tube sections of appropriate lengths and diameters, which fit into each other telescopically or which can be separated as required.

Considering that a maximum tolerance radiation dose is received from the radiation of a source of 10 mg-equiv of Ra at a distance of 1 meter over six hours, and that the radiation intensity falls in proportion to the square of the distance, it is easy to calculate the permissible working time with manual pneumatic holders of a definite length and a source of known activity. A manual pneumatic holder is designed for moving preparations from one shielded installation to another. Due to its simplicity, mobility and absence of a distributing mechanism, the pneumatic holder is appropriate for rapid operations and can be used successfully for work with high activity sources. Thus, for example, if 15 seconds are required to move a preparation with an activity of about 60 g-equiv of Ra from one container to another, then this may be accomplished with a 2-m pneumatic holder.

The length of the short arm of the main L-shaped tube of the manual pneumatic holder is chosen in relation to the depths of the cavities into which the holder is to be inserted and usually does not exceed 0.25-0.5 m.

The holder described does not hold articles without the operator's hand gripping the bulb, but with the use of a suitable closing device, it is readily converted into a self-holding instrument, although this somewhat complicates and slows operations.

In remote manipulation in shielded chambers, a need arises for holding hollow vessels, whose external diameters are either greater than the internal diameter of the pneumatic holder or when they are deeply ensconced in ring holders of laboratory stands, in ring sets for water and vapor baths, etc. Single-cavity holders cannot be used in these cases.

A double-cavity pneumatic holder makes it possible to hold articles by gripping them on the outside and by gripping them on the inside, thus differing from a single-cavity device in that the rubber sheath lines not only the inside surface of the bucket, but also the outside and forms in this way a second hermetically sealed cavity of variable diameter. Hollow articles are held by gripping them on the inside by inserting the pneumatic device into them to a certain depth and then filling the outer sheath of the pneumatic device with air.

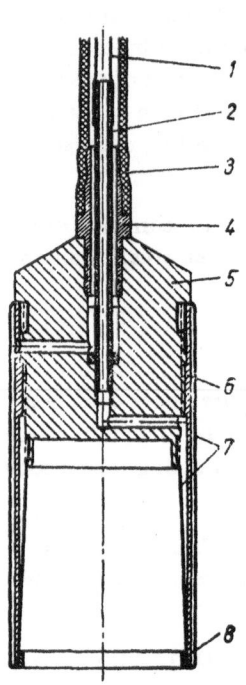

Fig. 11. Double-cavity pneumatic holding device: 1) inner pneumatic lead; 2) inner lead connection; 3) outer pneumatic lead; 4) outer pneumatic lead connection; 5) body; 6) bucket; 7) rubber sheath; 8) conical sealing ring.

The double-cavity pneumatic holder is illustrated in Fig. 11.

The two ends of the rubber sheath are sealed to the body with glue or by some other method.

A pneumatic holder of this design can only be assembled by the following sequence of assembly operations.

To a special groove in the lower end of the body is fixed one of the ends of the rubber sheath, which is smoothed out without the inner surface turned outward as shown in Fig. 11. The bucket is then screwed onto the body in such a way that the rubber sheath is inside it. The top end of the bucket does not yet close the second groove at the upper end of the body. The free part of the sheath, protruding out of the bucket, is then turned inside out over the bucket and smoothed out so that the sheath encompasses the walls of the bucket quite tightly, but not under too much tension. When the second end of the sheath has been fixed (tied with thread) to the body, a hose is attached to the outer lead, through which air is passed under pressure. The air compresses the part of the sheath inside the bucket and fills the outer part, pushing out a large part of the sheath from the bucket. When the sealing of the partly assembled pneumatic device has been tested, the bucket is screwed down manually to the limit without letting out the air from the inner cavity and holding the bucket from the outside together with the rubber.

When the bucket has been screwed down to the limit, the sealing ring is fitted into place, the hoses connected to the leads and the sealing of all the cavities and connections of the assembled parts tested.

Before the pneumatic holding device is completely assembled, its body is connected to the bucket and leads, whose threads are first greased with a thick and viscous lubricant (lanolin or vacuum grease), and the seals of the connections of the bucket and each lead to the body are tested.

A test of the seal showed that the pneumatic holding device retained articles for several weeks.

A comparison of the efficiency of attaching the rubber sheath to the body with thread or by gluing showed that the former method was the better.

The double-cavity pneumatic holding device described can perform double the number of operations that the single-cavity one can and its use would introduce a basic improvement in pneumatic manipulators.

Chapter IV

LIQUID DISPENSERS

1. Autopipette for Radiochemical Operations

As a rule, in work with normal pipettes the liquid is sucked in by the mouth of the operator or a plunger. The liquid is held in the pipette and run out of it by closing and opening the hole at the upper end with the finger and this is difficult in remote work and quite impossible without holding the pipette in the hand. In work with toxic and especially, radioactive and corrosive materials, such operating methods may result first, in poisoning and second, since very often the same volume of liquid needs to be measured a number of times, operations with a normal pipette are tedious and long.

In working with radioactive and corrosive materials, it is often necessary to carry out the operation at a distance, which is difficult to accomplish with normal pipettes (even with the use of certain accessory devices).

Pipettes fitted with a syringe-type head, used to eliminate the need for taking the end of the pipette into the mouth, have the following essential disadvantages: a) as the plunger of the syringe must be moved in and out, it is difficult or impossible to operate with one hand; b) due to the leakage of air through the gap between the plunger and the walls of the body, drops of liquid escape from the pipette and c) the portions of liquid must be constantly read off on the scale of the pipette, etc. Other devices added to pipettes are also not ideal and usually complicate and retard work, for example, a well-known type of fitting is a rubber bulb with an adjusting screw connected directly or by a hose to the pipette. Such a device complicates and increases the time required for washing the pipette during work with different liquids, in general, and radioactive ones, in particular.

The type of autopipette described is convenient and guarantees safe work.

The autopipette (Fig. 12) consists basically of the standard type pipette with an attachment for sucking in the liquid.

The soft-rubber connection joins the pipette to a standard type of rubber bulb hermetically. Operating the bulb sucks in, retains or expels liquid from the pipette. In order to stabilize the position, the bulb is enclosed in a body of two rigid sections — an upper and a lower.

Likewise, to give a more stable arrangement, the end of the pipette and other accessory attachments described below are also enclosed in the rigid tube of the connection.

Pressing the knob of the rod with the thumb causes the rod to squeeze the bulb as a preliminary to filling or to expelling liquid from the pipette. The bulb will return to its original shape (due to the elasticity of the walls of the bulb itself) and force back the rod, whose movement is strictly limited by an end stop in the upper section of the body.

An adjusting screw regulates the squeezing of the bulb and thus regulates the portion of liquid sucked in and expelled. A rough scale of the number of turns is first marked off on the rod during the setting up of the autopipette, which corresponds to the required volume. An exact setting is achieved by adjusting the regulating screw in relation to the readings of the pipette scale in preliminary tests.

Microregulation is achieved by squeezing the bulb after the dose of liquid has been collected in the pipette, by revolving the lower part of the body relative to the upper part.

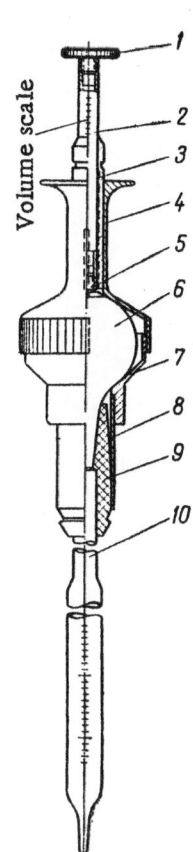

Fig. 12. Auto-pipette for radio-chemical work: 1) screw; 2) rod; 3) regulating screw; 4) top half of the body; 5) tip of rod; 6) bulb; 7) lower half of body; 8) connection; 9) collar; 10) pipette.

This arrangement makes it possible to measure automatically (without visual reading) a determined portion as well as arbitrary amounts of liquid. In other words, it is possible to perform all the operations carried out with a normal pipette both directly and remotely.

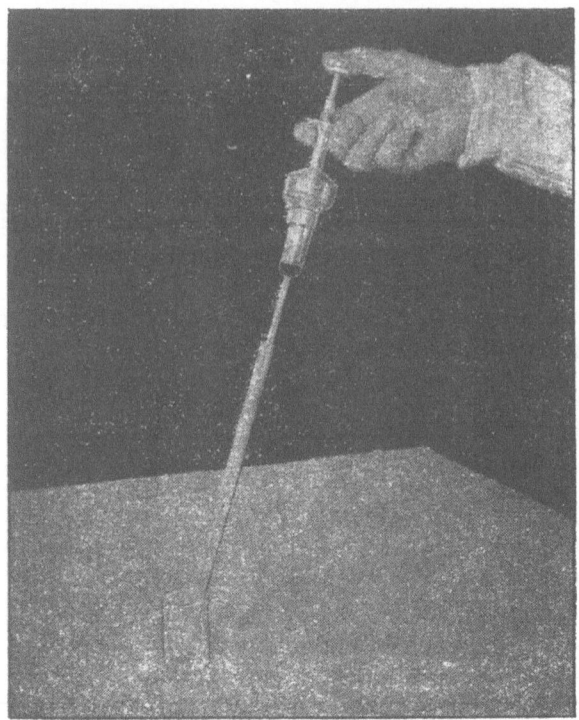

Fig. 13. An example of work with an autopipette. Collection of a liquid.

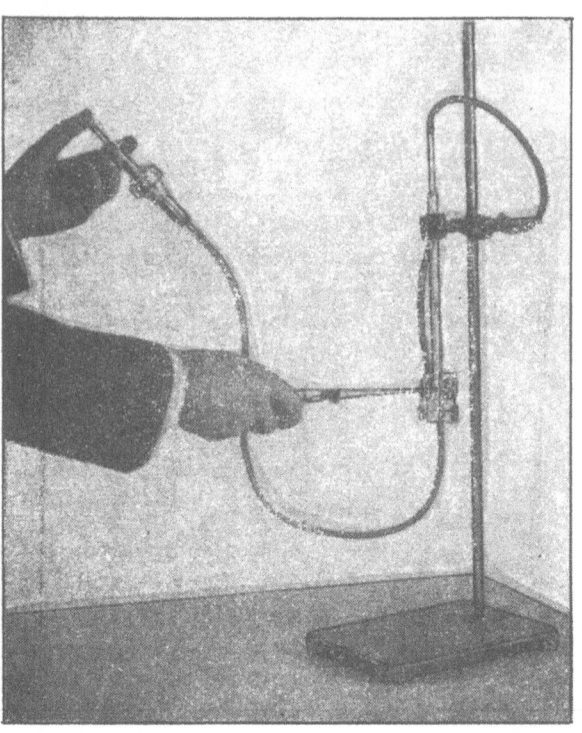

Fig. 14. An example of remote operation with an autopipette.

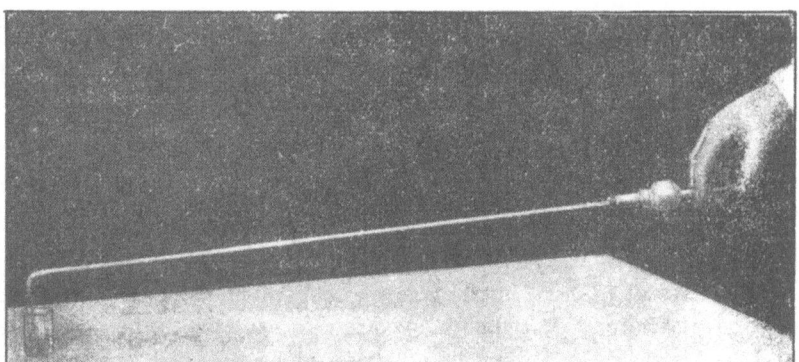

Fig. 15. Example of using an autopipette as a semi-remote manual sample collector.

An example of the use of an autopipette under the most typical conditions is illustrated in Fig. 13.

Remote operation with an autopipette is shown in Fig. 14. The pipette is fixed to a normal laboratory stand and connected to the attachment by a rubber hose and a glass tube of suitable diameter, with one end inserted into the plunger tip, instead of the shaft of the pipette, and the other into the connecting hose. The other end of the hose is fitted onto the shaft of the pipette.

Another method of using the autopipette as a sample collector is shown in Fig. 15, where a pipette of normal length or one that has been shortened to the necessary length (when used in difficultly accessible positions) is connected to the attachment by a light duralium L-shaped tube. The end of the long arm of the duralium tube is fitted into the tip (Fig. 12), while the pipette itself or (if no controlled reading is needed) a suitable glass or other tip is connected to the short arm of the tube with a rubber sleeve.

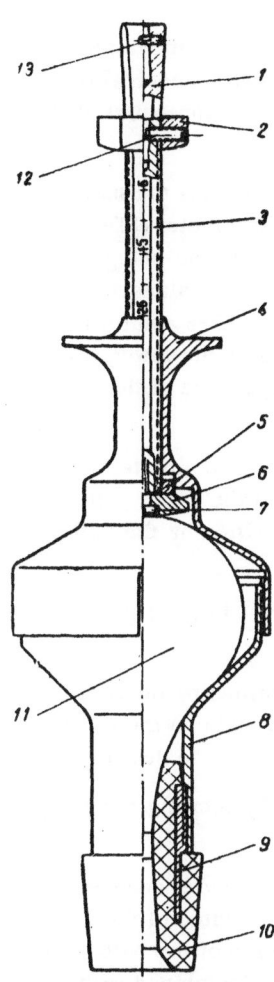

Fig. 16. Radiochemical hydrodispenser: 1) small head of plunger; 2) large head of plunger; 3) screw rod; 4) upper section of body; 5) nut; 6) clamping collar; 7) pivot screw; 8) lower section of body; 9) band of adaptor; 10) adaptor; 11) rubber bulb; 12,13) stop screws of heads.

The autopipette is designed for remote measurement of liquids by volume in the range 0.1-0.2 to 8-10 ml using normal chemical glass pipettes.

The set of attachments for measuring liquids consists of a selection of chemical pipettes of the required capacities, an L-shaped extension of duralium tube and a rubber hose with a tip. These make it possible to use the autopipette under any conditions and they are always to be found or may be readily made in any laboratory.

When it is necessary to work with the extension (for example, in removing radioactive liquid samples from a vessel in a container or in a shielded, exhausted cupboard) the glass pipette is attached to the short arm of the extension and the body of the controlling section to the long arm.

The pipette and the controlling attachment are connected by a rubber hose when the glass pipette is to be fixed in a stationary position at a distance from the operator, for example, to a stand in a shielded, exhausted cupboard.

In measuring arbitrary amounts of liquids, the scale on the glass pipette is used directly for reading off the volumes.

For measuring a large number of similar portions of liquids, the controlling arrangement may be regulated first for working with this volume, using the regulating screw, and the scale of the glass pipette, and taking off preliminary samples of liquid of similar physical properties.

The pipette is washed by repeated filling with and expulsion of washing solutions.

2. Radiochemical Hydrodispenser

The hydrodispenser is designed for remote measuring of radioactive or other harmful liquids by volume in single portions from 0.05 to 25 ml using the normal chemical or special interchangeable pipettes with pneumatic, manual or semiautomatic electromechanical control. The known devices for measuring various harmful liquids using a pneumatic method have a series of essential disadvantages.

In operating with them one cannot a) have a wide range of liquid aliquots, b) work without visual control, c) collect and pour off liquids at very different volume rates and d) use them alternately as manual or as stationary and as remote or semi-remote devices, etc.

The hydrodispenser described (Fig. 16) fulfills all the working conditions and is simple in construction and reliable and convenient in operation.

To set the original position and decrease the heat transmitted from the hands of the operator during operation, the rubber bulb of the hydrodispenser is enclosed in a rigid metal or plastic body, consisting of two sections, which screw together. The rigid body increases the accuracy of the measurements, protects the bulb from contamination and allows a more efficient arrangement of the bulb in relation to the other parts of the dispenser connected to it.

In the lower section of the body there is a thick-walled adaptor made of soft rubber and also enclosed in a rigid band. The shafts of normal or special pipettes are fitted into its opening. The adaptor's band holds the pipettes in place and ensures a hermetic seal with shafts of different diameter, the tips of pliable connecting hose of pneumatic leads and rigid tubes of extensions, of either straight or L-shape.

The bulb is squeezed before filling the pipette and to expel liquid from it by the screw rod, placed in the smooth-walled (without threads) cylindrical guide channel of the upper section of the body. The lower end of the rod is fitted with a clamping collar, rotating freely on a pivot screw. A scale is marked off on the central section of the screw rod for approximate dispensing and a thread is cut for exact dispensing. By revolving the

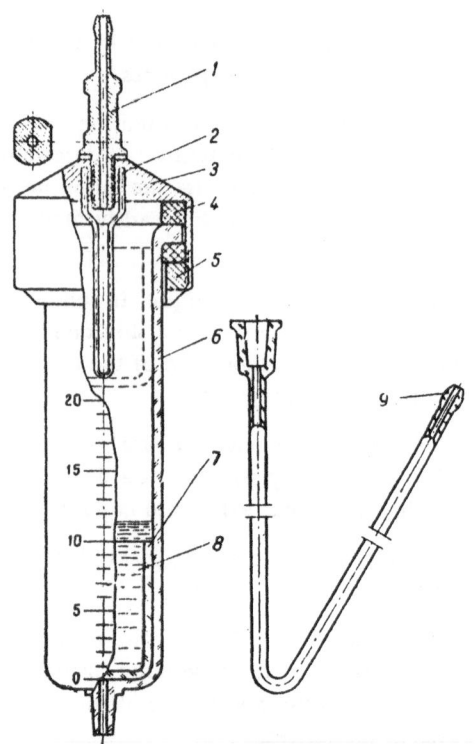

Fig. 17. Self-draining pneumatically controlled syringe-pipette: 1) ridged connection; 2) upper stop for plunger; 3) body; 4) packing ring; 5) screw ring; 6) body of the syringe; 7) syringe plunger; 8) wetting liquid; 9) angle tip for operations in a gas bath.

heads (the large or small one), it is possible to move the plunger relative to the frictional stop nut, when its conical outer surface enters the corresponding bore in the upper part of the body.

The small detachable head connects with the shaft of the reducing gear of a synchronous electric motor (SD-60) and is also used for rapid manual rotation of the plunger. The large head is designed for slow rotation of the plunger manually or by the synchronous electric motor SD-2 and for fixing the head with the stop screw. It also limits the lower position of the plunger by resting against the upper face of the upper section of the body.

The bulb is squeezed before filling the pipette, thus expelling liquid, by pushing in the plunger to the limit. The plunger is pushed back to its original position by the bulb resuming its former shape.

The size of the set volume is defined by the position of the stop screw.

An approximate volume is determined by the scale on the plunger and an accurate one by the scale on the pipette, during preliminary tests with liquids of suitable density.

The liquid is dispensed dropwise by rotating the plunger manually or with the electric motor.

As Fig. 16 shows, small-capacity pipettes connected to the dispenser decrease the overall dead volume of the system due to the decrease in the working volume of the bulb, which results in greater accuracy in dispensing (combined with a wide range of aliquots and volume rates) and the possibility of using standard electric leads for automatic working conditions.

3. Self-Draining, Pneumatically Operated Syringe-Pipette

The use of various supplementary devices makes it possible to use normal chemical pipettes for measuring toxic liquids. However, in work with γ-active and chemically aggressive liquids, such devices do not ensure protection from their effects.

The syringe-pipette described (Fig. 17) may be used for direct and for remote work with toxic liquids and gases. It has a combined solid-liquid plunger and a self-draining tip. The absence of dead space makes it possible to work with gases.

The combined plunger ensures practically complete impermeability of the gaps between it and the body, and great mobility is achieved by forcibly filling the gaps with a mobile, neutral liquid, e.g., distilled water. This neutralizes molecular forces which act on the plunger when the gaps are partially filled with liquid and make it impossible to move it smoothly by pneumatic means.

The self-draining end of the outlet tip of the syringe-pipette operates due to the fact that the pressure above the plunger is always slightly less than that below it, so that the last drops of liquid do not fall off the end of the outlet tip but are sucked back into the pipette. This eliminates specific contamination of the working area and prevents loss of the product.

The impermeability of the combined plunger and absence of dead space make it possible to use the syringe-pipette for remote measurements of gases.

The syringe-pipette is best controlled by a hydrodispenser, hydromanipulator or pneumatic hydromanipulator. In individual cases, other methods of pneumatic control may be used.

Standard solid-walled syringes of the required capacity and hollow plungers, which are shortened to the required length, may be used for preparing syringe-pipettes.

The connection of the syringe is ridged to make it more convenient for hermetically connecting rubber pneumatic leads with different sized bores. A stop limiting the lift of the plunger is made so as to prevent the liquid of the plunger flowing into the controlling pneumatic system. The body may be made of plastic or metal.

The surfaces of the lower packing washer and the screw ring, which bear on each other, should be powdered with talc. The plunger of the syringe, flooded with liquid, should slide smoothly down the body under its own weight.

The first filling and subsequent changes of the plunger liquid, without dismantling the syringe-pipette, are most conveniently accomplished by inverting the syringe and filling it with a normal small rubber pear-shaped bulb (for syringing) which is connected to the glass tip of the syringe.

An angled tip is used for work with gases in water baths. Detachable sockets are used for connecting the syringe-pipette to the triple-cavity pneumatic holders of pneumatic hydromanipulators.

4. Radiochemical Hydrocolumn

In working with liquids, it is often necessary to measure and pour them from one vessel to another using remote devices, most of which are not safe or convenient for work.

The devices described below for transferring and volumetric measurement of toxic liquids make it possible to perform these operations, which are usually interconnected, together, even when the working vessels are at different levels and the one to be filled is at a much higher level than the one from which it is filled. This equipment automatically establishes a constant starting liquid level in the supplementary (measuring) volume, which is often required practically in transferring liquid from one vessel to another in definite portions.

The devices developed consist of two complementary volumes, connected by an air lead — a controllable, measuring volume of constant capacity for liquid and a controlling volume of variable capacity for air. The system has two liquid leads (to suck in and to expel), with inlet and outlet valves, which are loose in their seats.

The valves close the appropriate channels at the required moment under their own weight.

The connections and details of a radiochemical hydrocolumn, to be filled from above, are illustrated in Fig. 18. The inlet tip with the seat for the inlet valve is connected to the inlet tube with a ground-glass joint. The inlet tube is also connected to the measuring cylinder by a ground-glass joint, in the form of a hollow stopper for the measuring cylinder. These connections seal the system hermetically.

The inner cavity of the stopper acts as a reservoir for collecting liquid if the measuring cylinder is accidentally overfilled; the stopper is then slowly filled through an eccentrically placed opening, which lets in air under normal operation.

The measuring cylinder terminates at the lower end in an outlet tube, which is bent up with a seat for an outlet valve. With this arrangement, the valves operate without additional devices, simply by their own weight.

The outlet tip is also connected to the measuring cylinder by a ground-glass joint.

A flexible hose connects an appropriate lead on the measuring cylinder's stopper to a T-joint and an elastic bulb. In order to establish the initial level of the liquid in the measuring cylinder and then seal off the air-cavity system after regulation, before the beginning of operation the screw of the T-joint is used to open the air-cavity system to the atmosphere. Instead of the screw, another type of device may be used, e.g., a piece of rubber tubing with a screw or spring clip, a three-way tap, etc.

All the devices, including the inlet and outlet valves, are best made wholly out of heat-resistant glass as this material is very stable to the action of chemically active and radioactive liquids and vapor. The most convenient form for the outlet and inlet valves is a cone with a hemispherical top and bottom.

Experiment has shown that in working with liquids whose density is close to unity, the valves are best made in one piece from, for example, normal glass rod or thick-walled pressure or capillary tubing. If an air cavity is planned in the valves, then it should be small so as not to decrease their overall weight and tight fit in their seats (or even allow their floating in the working medium).

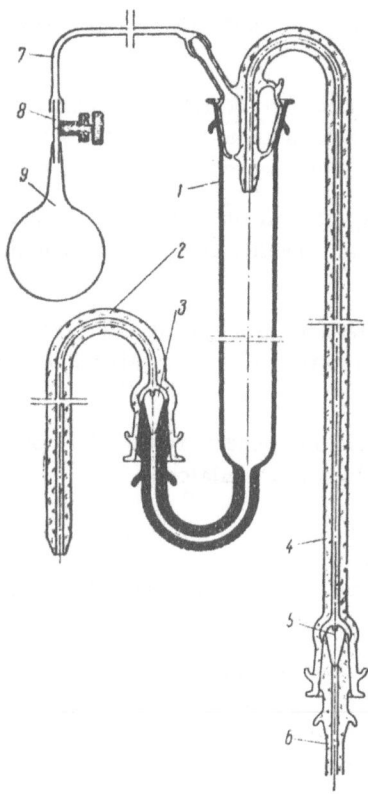 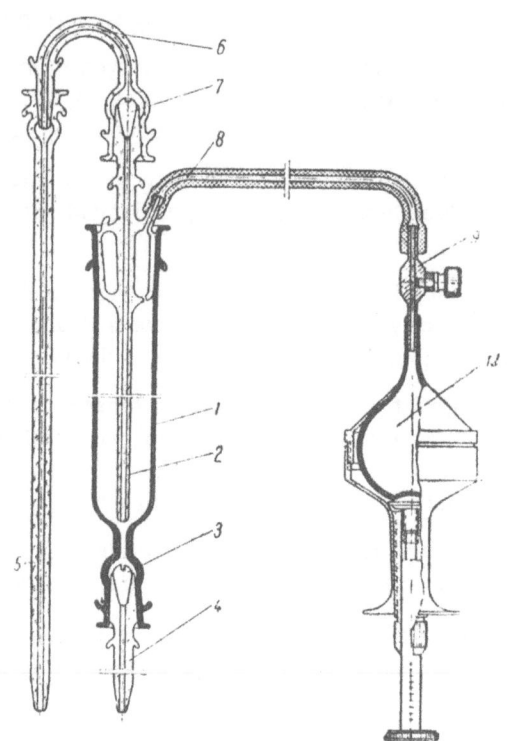

Fig. 18. Radiochemical hydrocolumn filled from the top: 1) measuring cylinder; 2) outlet tip; 3) outlet valve; 4) inlet tube; 5) inlet valve; 6) inlet tip; 7) connecting hose; 8) T-joint with screw; 9) elastic bulb.

Fig. 19. Radiochemical hydrocolumn, filled from the bottom: 1) measuring cylinder; 2) inner inlet tube; 3) inlet valve; 4) inlet tip; 5) outlet tip; 6) connecting elbow; 7) outlet valve; 8) connecting hose; 9) T-joint with screw; 10) controlling mechanism of an elastic bulb.

For work with liquid media of high density (close to that of normal glass) the valves must be made of more heavy glass, for example, lead glass, or of normal glass, but with a heavy metal (steel, brass, tungsten, etc.) cone sealed inside it.

To eliminate the possibility of the valves sticking in their seats, the apex angle of the cone should not be too small. Theoretically, it should be greater than the angle of friction of glass on glass with roughly polished surfaces. Glass valves with an apex angle of the cone of about 30° operate well.

The working diameter and height of the measuring cylinder are selected in accordance with the required maximum volume and accuracy of measurement; the length and form of the inlet and outlet tips are chosen in accordance with actual working conditions, in particular, with the form, height and level of the vessels used.

In order to adapt the device for various working conditions, it must have a set of interchangeable inlet and outlet tips of various dimensions. They may be made of separate parts connected by ground-glass joints and this makes it possible to work with various types of vessels (beakers, dishes, narrow-necked and tall flasks, etc.) at different levels. The whole device may be used as a remotely controlled pump for pumping liquid working media from one vessel into another and simultaneously measuring the amount of liquid, which is most necessary in remote work.

When the vessel emptied is situated above the vessel being filled, the liquid may be syphoned naturally with remote control for only the start and the end of the process by the pneumatic controlling device.

Figure 19 shows a hydraulic column that is filled from the bottom, which is designed for use as a remotely operated buret, pump and syphon. Its arrangement and operation are analogous to the column described above, which is filled from the top, and no further explanation is needed.

Three types of device and mechanism may be used for remote pneumatic control of the column:

1. A rubber bulb with no set position for pumping and measuring arbitrary portions of liquid by the scale on the column. The device makes it possible to set the starting level of the liquid in the column automatically after each successive portion has been removed (Fig. 18).

2. A mechanism with an elastic bulb, fixed in a metal body, similar to the one used in an autopipette. This bulb may be used for automatic dispensing of liquid aliquots, first established by the scale of the column (Fig. 19).

3. A controlling mechanism including a pump plunger and a combined syringe (of the "Record" type), with a screw rod and other devices. This mechanism may be used for automatic and combined regulation of the rates of filling and emptying of the columns.

5. Micropump for Radiochemical Micropipettes

Volumetric measurement of radioactive and other harmful liquids in the range 0.001 to 0.2 ml with normal glass micropipettes of 0.1 and 0.2 ml capacity is difficult as the traditional methods do not satisfy the elementary safety requirements.

The set of devices for measuring microvolumes of liquids include two interchangeable standard-type glass micropipettes with capacities of 0.1 and 0.2 ml and a micropump, consisting of a metal body with a screw plunger, a rubber packing collar and a tightening nut (Fig. 20).

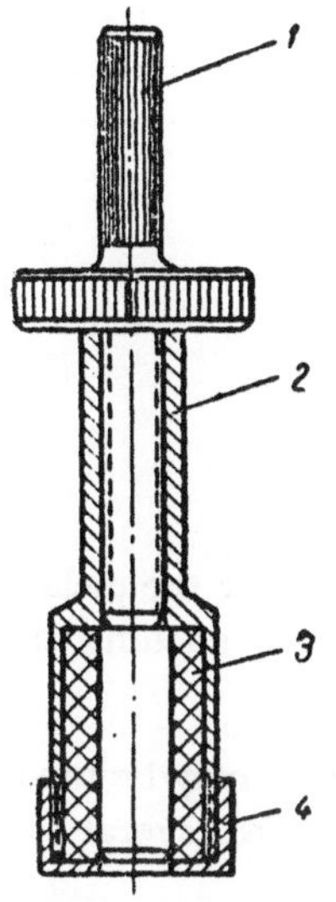

Fig. 20. Micropump for radiochemical pipettes: 1) screw plunger; 2) metal body; 3) rubber packing collar; 4) tightening nut.

The screw plunger is equipped with two heads of different dimensions for rotation at different rates. For rapid drawing in and expulsion of liquids from the micropipettes, the small-diameter head is rotated, and for exact correspondence of the liquid meniscus with the divisions of the scale, the large-diameter head is used.

The ends of the micropipettes are inserted into (or withdrawn from) the rubber packing collar of the body by screwing (or unscrewing) the tightening nut on the micropump body to ensure a hermetic seal of the micropipette with the micropump.

The body and screw plunger of the micropump are made of any type of wear-resistant material that fits together well and, as a rule, this is metal that is stable to corrosion or plated with an anticorrosive and antiwear coating (for example, chromium-plated brass). The packing collar is made of soft but dense (vacuum-type) rubber, for example, suitable sections of vacuum-rubber hose.

The micropumps and micropipettes are freed from specific contamination after operation by the normal methods (preferably combined hydromechanical ones) with special washing media. The micropumps and micropipettes should be washed free from radioactive contaminants in particular, first while still assembled in the working area, and then when dismantled.

The simplicity of the micropump construction enables the operator to take it apart and reassemble it in the working area.

After prolonged practical testing, the micropump has been used not only under the specific conditions for which it was initially designed, but also for work of a preparative chemical nature in normal laboratory production conditions with harmless liquids, and this illustrates its practical value.

6. Foot Micropump

A foot-force pump with a reservoir of variable capacity, made from an extendable and elastic material, may be used for various physical, chemical, medical, technical and other work connected with the measurement of normal and harmful liquids, using pipettes, capillaries, blood-mixing pipettes and other devices.

Tests with the foot micropump under laboratory and working conditions have shown it to be convenient and to ensure safe operation (Fig. 21).

The pump consists of a nozzle with a clip and a pedal, which rocks on its axis under the pressure of the operator's foot.

The pipette, blood mixer or capillary is attached by a rubber connector to the nozzle, which has a radial opening (for letting air in and out), opened and closed by the operator's thumb. The nozzle fits onto one end of a hose and the pump hose fits into the other.

The pump hose is placed in grooves in the base and is squeezed by the operator's foot by means of the pedal.

For convenience, the nozzle is fitted with a clip (which is placed on one or two fingers) or with a flat plate for directing the opening in the way required. Using the clip or plate, the operator can place the pipette, capillary or blood-mixing pipette in the required position.

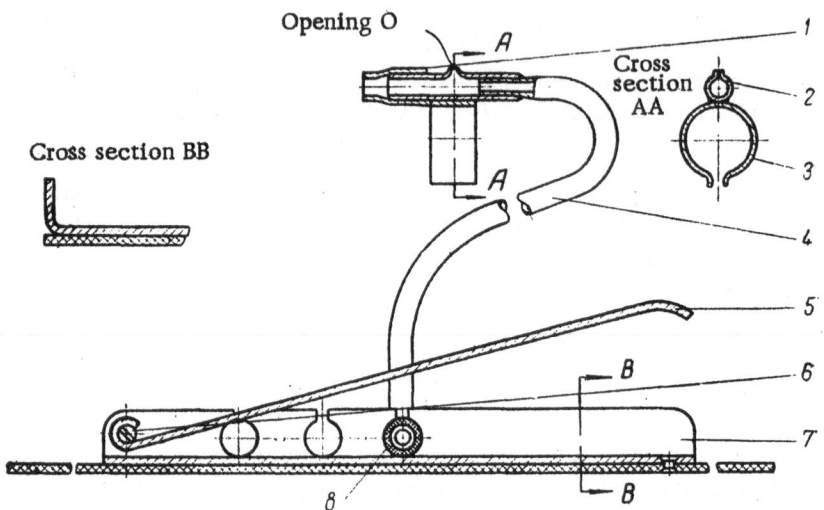

Fig. 21. Foot micropump: 1) rubber connector; 2) nozzle; 3) clip; 4) rubber hose; 5) pedal; 6) pedal spindle; 7) pump base; 8) pump hose.

In order to expel the liquid from capillary vessels, it is necessary to close the opening and press the pedal with the foot. The liquid may be expelled from noncapillary vessels either by using the pedal or by opening the nozzle.

A straight-pump hose may be used for volumes up to 1-1.5 ml; a U- or S-shaped pump hose may be used for volumes up to several milliliters. A rubber bulb fixed under the pedal is best used for volumes of several tens of cubic centimeters.

Chapter V

RADIOCHEMICAL HYDROMANIPULATORS

In many cases, the protective facilities used for work with radioactive materials must be applied in complex static and dynamic combinations; their character, relative spatial position and sequence of use, determined by the technical processes, often create great difficulties in operation.

All this makes it necessary to develop simple, but multifunctional devices and equipment that are interrelated to a minimum for remote operations.

The use of the hydromanipulators described makes it possible to carry out technical operations of measurement and transference to any point in the working space of portions of liquid from 0.01 to 20 ml, titration, and filtration of solutions, dispensing of different volumes, including into narrow-necked ampoules, etc.

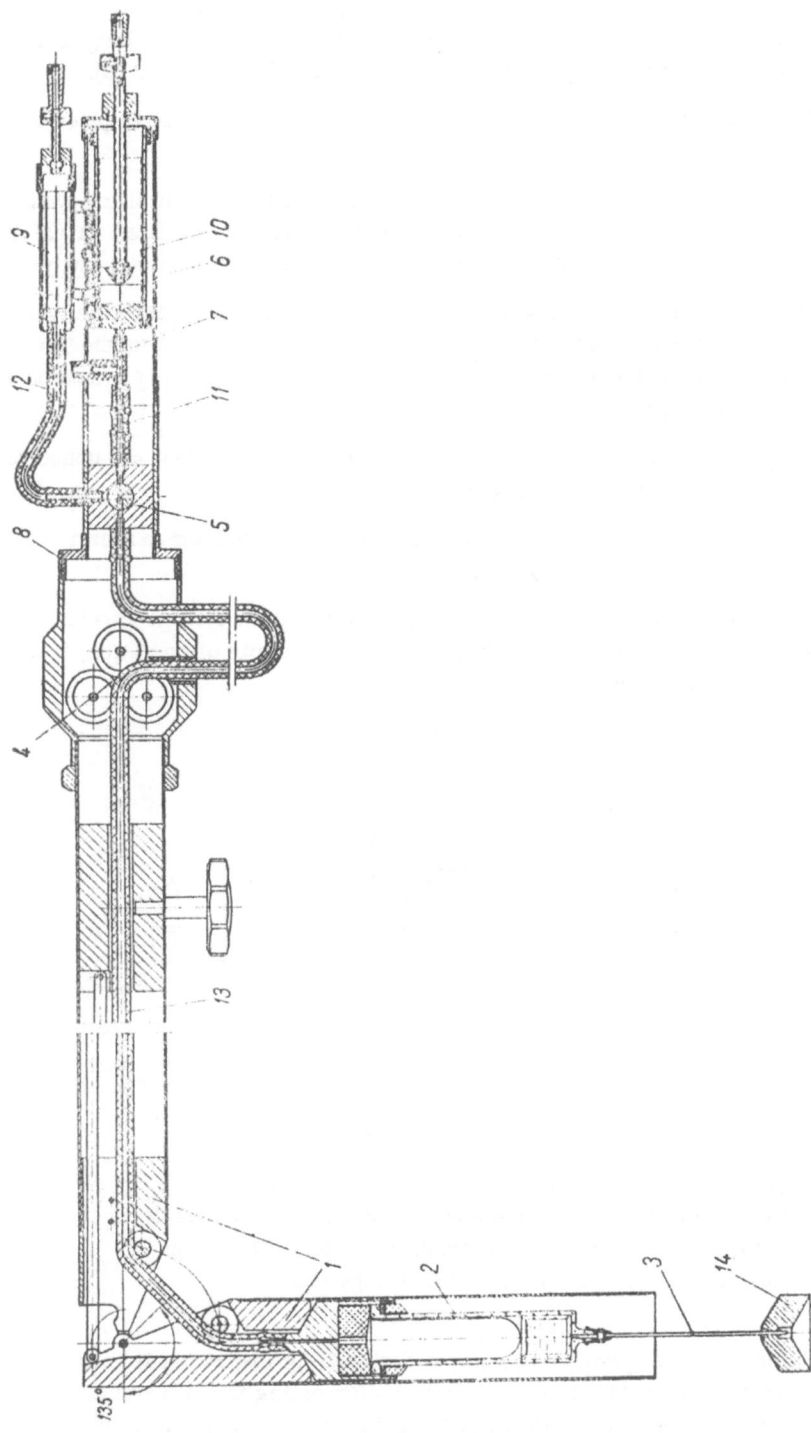

Fig. 22. Stationary manual hydromanipulator: 1) hinged tubular mechanism; 2) syringe-pipette; 3) needle tip with center-ing device; 4) hose-extension roller mechanism; 5) three-way distributing tap; 6) dispensing two-syringe panel; 7) T-joint; 8) connector; 9) dispensing syringe with 1-ml capacity; 10) dispensing syringe with 20-ml capacity; 11, 12 and 13) connect-ing hoses; 14) centering device.

47

1. Stationary Manual Hydromanipulator

A stationary manual hydromanipulator is illustrated in Figs. 22-24. Together with the syringe-pipette, it forms a pneumatic hydraulic mechanism for measuring and dispensing.

The manipulator may be moved along its axis to cover the working space; the main horizontal tube rotates around its axis in a fixing and guiding collar. In addition, the angle of the elbow of the hinged mechanism may be varied and the syringe-pipette moved vertically by a hose-extension mechanism (Fig. 22).

The plunger and the separating membrane (Fig. 24) of the syringe-pipette are not used in pneumatic pipetting of the working media.

Hydraulic pipetting of liquids is carried out by first filling all the measuring and controlling systems of the hydromanipulator above the plunger of the syringe-pipette with an inert liquid, usually distilled water. It is used primarily in work with back pressure and with chemically aggressive and volatile liquids, especially γ-active ones.

In pipetting by the pneumatic method and in reading off volumes on the normal scales of the syringes of the dispensing panel, the liquid is measured mainly by running off successive aliquots. This method gives greater accuracy of measurement and a result, which is independent of the density of the liquid being measured and the length of the inlet-outlet tip of the syringe-pipette (Fig. 22).

The sections and parts of the hydromanipulator are made of heavy or light metals that are noncorroding or plated with anticorrosion materials, glass, rubber, plastics, etc.

The tubular hinged mechanism is used for housing the other parts of the manipulator as shown in Fig. 22 and is exactly like the mechanism described in Chapter III, Section 1.

Hose-extension roller mechanism (Fig. 22) is needed for moving and fixing the syringe-pipette vertically by changing the length of the flexible connecting hose. In construction and principle of operation, it is similar to the mechanism described in Chapter III, Section 1.

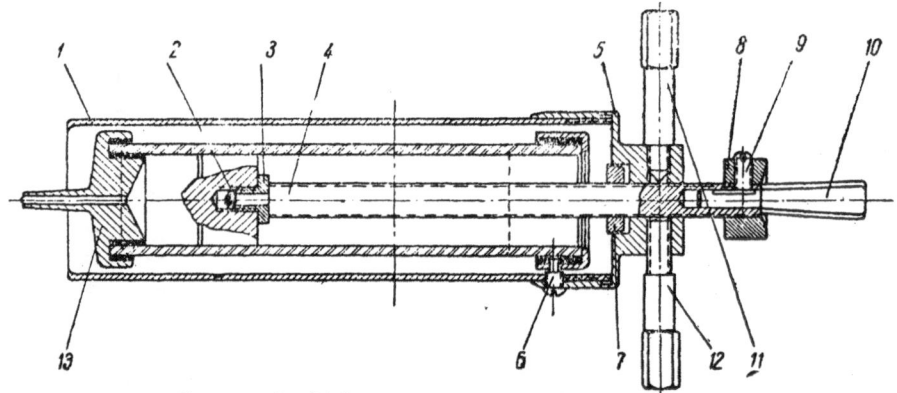

Fig. 23. Dispensing syringe with a capacity of 20 ml: 1) metal housing; 2 and 3) parts holding the rod in the syringe plunger; 4) screw rod; 5) head of housing; 6) pivot screw; 7) stop nut; 8) large head; 9) stop screw of large head; 10) small head; 11 and 12) stop screws; 13) syringe tip.

The pump-distributing mechanism of the hydromanipulator is designed for dispensing liquids. The mechanism consists of a dispensing two-syringe panel (Fig. 22), a three-way distributing tap, a T-joint, a connector and connecting hoses. The dispensing panel consists of two dispensing syringes with capacities of 1 and 20 ml, respectively. The scale of the small syringe should be marked off in units of 0.02 or 0.05 ml and that of the large one in 1- or 2-ml units.

The metal tube housings of the dispensing syringes are placed one above the other in parallel and are connected by screws and clamping collars. The large dispensing syringe is designed for measuring 1 to 20 ml of liquid, the smaller one for from 0.01 to 1 ml; liquids may be measured in drops with the smaller syringe.

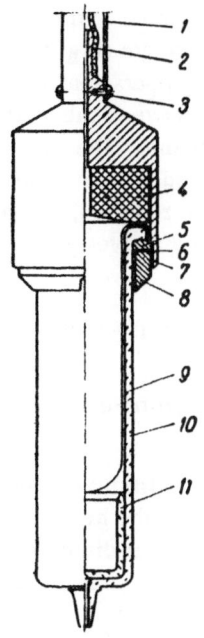

Fig. 24. Syringe-pipette of stationary manual manipulator: 1) L-shaped tube, which encloses the syringe-pipette (only in the dispenser and sample collector); 2) rubber hose; 3) screw; 4 and 5) rubber lining; 6) washer; 7) head of syringe-pipette; 8) threaded ring; 9) separating membrane; 10) syringe body; 11) plunger.

The dispensing syringe (Fig. 23) with a capacity of 20 ml is a combined glass and metal syringe of the "Record" type, which, in contrast to the normal medical injection syringe, is fitted with a rotating screw rod, a cone-shaped friction stop nut, stop screws and two controlling heads. It is designed for measuring from 1 to 20 ml of liquid.

The syringe is enclosed in a metal tube housing with a head and is prevented from rotating around its axis and moving along it by a pivot screw, in such a way as to leave some possibility of rocking and moving in relation to the housing tube and its screw. This fitting of the syringe is necessary due to the slight eccentricity of the inner section of the glass cylinder relative to the outer action of the metal casing. The screw rod allows rapid advance of the plunger and slow micromovement for greater convenience in particular technical operations. For this purpose, the opening in the head is smooth (without threads). The stop screw has a conical end, whose apical angle equals that of the cross section of the thread on the rod. The axial movement of the screw rod is stopped by tightening the screw to the limit. This makes it possible to move the plunger of the syringe micrometrically by rotating the rod manually. Unscrewing the screw slightly allows the rod to be moved unhindered along its axis. So that the plunger of the syringe will not move by itself during operation, the rod's movement is limited by a stop screw.

The stop screws are used in operating the syringe with one hand, leaving the other free for other associated operations.

The conically shaped nut limits the movement of the rod to the right. When the rod is rotated and pulled simultaneously to the right, the nut, previously set on the screw rod, wedges its outer conical surface into the bore of the head and rigidly limits the travel of the rod and of the syringe plunger attached to it. Hence, liquid portions, previously set by the scale, can be measured and the need for visual control and the possibility of mistakes are eliminated. This method is particularly convenient for measuring a large number of similar samples.

The large head is used for slow rotation of the rod in manual work and the small one for rapid rotation. It is readily demountable so that it can be removed when the syringe pump is set up for the automatic introduction of liquids and gases by connecting the rod to the shaft of a synchronous SD-2 electric motor (Warren motor).

In pneumatic and hydraulic operation, the plunger and inner surface of the glass cylinder are best lubricated with lanolin. Before being set in the dispensing panel of the hydromanipulator, the standard syringe is lubricated and the balance and ease of movement of the plunger in the glass cylinder and the relation of the total length of the volume scale to the actual diameter of the plunger are tested.

The dispensing syringe with a capacity of 1 ml is similar in construction and operation to the large dispensing syringe and all that has been said above applies to it as well, so that no further explanations are needed.

The three-way distributing tap (Fig. 22) is designed for connecting the syringe-pipettes to the dispensing syringes either in turn or simultaneously.

The T-joint (Fig. 22) connects the pneumatic hydraulic system of the manipulator to the atmosphere and external vessels and, in the absence of the three-way distributing tap, to the dispensing syringes. The T-joint is not an essential part of the hydromanipulator.

The connector (Fig. 22) connects the hose-extension and the pump-distributing mechanisms of the hydromanipulator, preferably with a "covering" nut. The same type of nut is best used for connecting the hinged mechanism to the hose-extension one.

The connecting hoses (Fig. 22) must be thick-walled and rigid so as to eliminate the possibility of sharp bends in them and excessive extension during operation. The bore of the hose and the thickness of the walls should be about 2 mm.

The syringe-pipette is designed for transferring liquid working media from one vessel to another within the working space covered by the hydromanipulator and is a solid-walled syringe with a shortened plunger (Fig. 24). The body of the syringe is fixed in the head in a rubber lining with a nut. In hydraulic pipetting operations with chemically aggressive and with γ-active liquids, a movable, impermeable separator is placed between the working and the controlling media, taking the form of a rubber or plastic, smooth or crimped membrane. The plunger and the membrane are not used in pneumatic pipetting. The body of the syringe may be replaced by a suitable pipette (for example, a Mohr pipette). In special cases, when transparency of the pipette is not required, it may be made of a material other than glass. For example, in working with hydrofluoric acid, which dissolves glass, the pipette may be made of ebonite, a material stable to the action of this acid.

When the main sections of the hydromanipulator are used in a manual, angled dispenser (sample collector) a screw is used for attaching the syringe pipette to the short arm of the tube.

A hose is used as a pneumatic or a hydraulic lead and connects the syringe-pipette to the pump-distributing mechanism of the hydromanipulator.

In certain cases, for example, when the syringe-pipette must be moved a great distance vertically without swinging freely on the connecting hose in its lower position, an extension tube may be attached to the head of the syringe-pipette. This tube must have a bore equal to the outer diameter of the head and its outer diameter must be slightly less than the inner diameter of the short arm of the main tube of the manipulator so that the extension tube can be moved freely and telescopically in the short arm of the main tube.

The inlet and outlet needle tip with the centering device (Fig. 22) is designed for connecting the syringe-pipette to the vessels containing the working media. This device is used for work with narrow-necked ampoules, dispensing and dissolving in them powdered and baked radioactive preparations.

The centering device automatically centers the tip relative to the neck of the vessel. When the syringe-pipette is lowered onto the vessel under its own weight, the conical cavity in the working end of the centering device takes up the required position on the neck of the vessel.

The inlet-outlet tips used for the syringe-pipettes are standard syringe needles of various diameters and lengths, made of stainless steel, Bogush needles or are of glass or other material.

2. Multichannel Hydromanipulator

Remote manipulation of laboratory chemical vessels and apparatuses greatly hinders elementary operations with liquid media (pouring from one vessel into another, dispensing, mixing, filtering, titrating, collecting samples, etc.).

The device proposed for remote manipulation of harmful liquids ensures the safe accomplishment of the above operations and a series of others with a number of liquid media simultaneously; it also allows the injection and removal of biological liquids from experimental animals.

This method may be used in work with gaseous media. It consists of using a pump and plunger multichannel measuring system with a solid-liquid plunger, which arbitrarily changes accessory space for the working media. The measuring system is connected to the main cavities by a lead with distributing devices and additional devices for carrying out the necessary technical operations. The system provides for remote control of the movement of working media in the necessary sequence by moving the solid-liquid plunger.

The construction of a multichannel hydromanipulator is illustrated in Fig. 25.

The accessory space for working media is separated by a movable barrier (for example, a piston or an elastic membrane) into two cavities — a controlling one and a controlled one. The measuring and controlling cylinders of various volumes, 2, 3 and 4, are connected together and to the controlling cavity of the accessory space with a four-way connector and thick-walled, flexible hoses, forming a closed measuring and controlling hydraulic system of constant volume. In the working state, the system is filled completely with a neutral liquid (for example, distilled water).

The controlled cavity of the accessory space is connected through a distributing apparatus with one or several multi-way taps to a system of sucking and blowing tubular leads and accessory devices, which form together a remotely controlled pumping and measuring hydropneumatic system for working media.

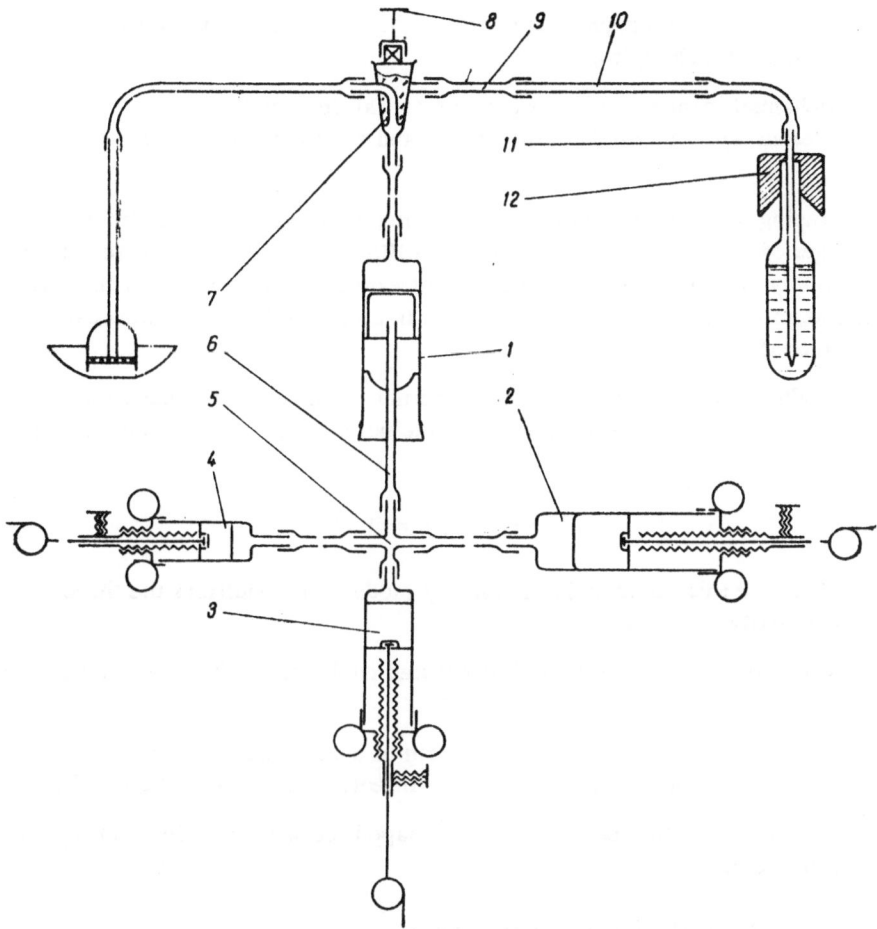

Fig. 25. Plan of multichannel hydromanipulator: 1) accessory space; 2) large control cylinder; 3) medium control cylinder; 4) small control cylinder; 5) four-way connector; 6) flexible hose; 7) distributor; 8) distributor key; 9) lead connector; 10) observation tube; 11) tip of syringe needle; 12) centering device.

The number of channels of the distributing apparatus and the number of tubular leads and accessory devices connected to them are determined by the number of vessels with working media and the number of technological operations. Vessels with working media may be placed on a lifting turntable.

For work with comparatively small amounts of liquid (e.g., several liters), the measuring and controlling cylinders can be glass medical syringes of 100- and 10-ml capacity and the so-called tuberculin syringes of 1-ml capacity. The accessory space may be a Liuer's syringe, made entirely of heat-resistant glass. The hollow plunger of this syringe is shortened (cut) to the required length and the open end of the body closed with a plug (e.g., a rubber bung).

The mechanism for hand movement of the plunger must provide for both forward and reverse and screw motion of the plunger rods. The range of movement may be limited with stops and stop screws. The need for making observation tubes and glass connecting tubes is then eliminated. The key of the distributor may be at a distance, i.e., connected with a rod of the required length. The centering device is a weight, moving freely down the needle-outlet tip to a shoulder and, in this position, centers narrow necks of vessels relative to the tip. The apparatus described makes it possible to work with harmful liquids and gases at a distance.

The pump-plunger measuring system with a combined solid-liquid plunger makes it possible to charge arbitrarily the volume of the accessory space, allowing remote control and dispensing transfers of working media in the required order.

Working media are removed from appropriate parts of the system by pumping liquid into the controlling cavity of the accessory space (or sucking it out).

By means of multichannel manipulators it is possible to transfer working media from one vessel to another, to dispense liquids into vessels with extremely narrow necks and to inject liquids and gases into experimental animals.

Measuring and pumping different volumes of working media are accomplished by moving liquid in the hydrosystem with the plunger of the three measuring and controlling cylinders. Their mechanism makes it possible to pump and measure rapidly volumes of working media, not exceeding the volume of the operating cylinder, to fill the cylinder with determined doses or to measure liquid in drops. All the apparatus can be washed readily or flushed both during work or afterward.

The number of possible combinations of remote movement of media from one attached vessel to another, without the use of additional manipulative devices equals the number of all the possible arrangements of n elements, i.e., it equals

$$1 \cdot 2 \cdot 3 \ldots n = n!$$

For work with radioactive substances under laboratory production conditions the three following basic appliances are required practically:

1. A stationary hydromanipulator for shielded, exhausted cupboards with a multichannel distributing apparatus.

2. A transportable hydromanipulator, made in the form of a container, with an internally arranged controlled accessory space and distributor, mainly intended for injecting experimental animals.

3. A hand hydromanipulator in the form of a long L-shaped rod without a distributing apparatus for withdrawals, transfers, sampling, etc.

3. Other Modifications of Hydromanipulators

The stationary hydromanipulator consists of the measuring-dispensing hydropneumatic mechanism of the stationary manual hydromanipulator described, i.e., its pump-distributing part and syringe-pipette combined with a U-shaped manipulator of the "mechanical hand" type.

To produce such a combined apparatus, a syringe-pipette is attached to the controlled part of a mechanical or other manipulator, and the pump-distributing mechanism to the controlling part of the main manipulator in a special holder, or left lying free or hanging from it, so that it may be held in the hands in subsequent work.

Considering the working conditions of radiochemists and the need for visual control of processes in shielded, exhausted cupboards, it must be admitted that the method of fixing the syringe-pipette in the holder of the main manipulator and the free positioning of the pump-distributing mechanism on the opposite side of the shielding screen are most maneuverable and best satisfy the given working conditions.

The hose connecting the syringe-pipette to the pump-distributing mechanism may be passed through special openings and channels in the main manipulator or shielding wall.

The hydrodispenser for stationary work is the same measuring and dispensing mechanism of the stationary manual hydromanipulator, but with fixed attachment of the syringe-pipette. Such a hydrodispenser is used for measuring liquids volumetrically without moving them.

The introduction of liquids and gases by automatic means may be provided for by equipping the dispensing panel of the hydrodispenser with an electromechanical drive by connecting in a synchronous electric motor of the SD-2 or SD-60 type.

The manual angled dispenser and sample collector is designed for subsidiary operations, mostly performed under open conditions.

The operator is shielded from the harmful effect of radiation by distance and the short duration of the operation carried out, as well as by the use of additional protective facilities (stationary and mobile shields, containers, etc.).

The dispenser and sample collector consists of a rigid L-shaped tube with arms of different lengths, with a syringe-pipette attached to the short arm by the method described in Chapter V, Section 1 and to the open section of the long arm, a dispensing panel with a distributing tap or T-joint or only one syringe dispenser of the required capacity. The connecting hose is fitted inside the L-shaped tube.

The length of the long arm of the main L-shaped tube is selected in relation to the actual working conditions. The most convenient and practical length of tube for the work is up to 2-3 m. In order to decrease the weight of the tube (without decreasing its strength and reducing the length) in certain cases it is designed and constructed as a cantilever of approximately equal bending strength, from tube sections of appropriate lengths and diameters, which fit into each other telescopically or which can be separated as required.

The length of the short arm of the main L-shaped tube is chosen in relation to the depths of the cavities into which the syringe-pipette is to be inserted. For work under normal conditions, the length of the short arm does not exceed 0.1-0.25 m, but it can be longer in special cases.

The use of a needle inlet-outlet tip with a centering device on the syringe-pipette of a dispenser and sample collector allows the operator to dispense radioactive liquids into narrow-necked vessels, placed in the depth of shielding containers that are not within the direct radiation zone. The simplicity and maneuverability of the dispenser and sample collector makes it possible to use it for operations with radiation sources of relatively high activity.

Chapter VI

RADIOPREPARATIVE PNEUMATIC HYDROMANIPULATOR

The pneumatic manipulative device is used for remote work with γ-active liquids in shielded cupboards.

The bases of the pneumatic hydromanipulator are the previously developed and tested pneumatic manipulator and hydromanipulator, which are combined in one multifunctional apparatus. This combination became possible after the creation of double- and triple-cavity pneumatic holders and the pneumatic hydrodispenser.

The combined pneumatic hydromanipulator will transfer primary and transportation containers with radioactive preparations, in particular, average- and low-capacity chemical laboratory glassware and ampoules of the most diverse types. With the manipulator it is possible to measure volumes of from 0.01 to 100 ml and to transfer liquids (including low-boiling radioactive, chemically aggressive, glass-dissolving ones, etc.) remotely, using readily changeable pipettes.

The manipulator will introduce accurate portions of liquids and gases into any part of the working space from the outside and dispense into glass ampoules of various forms.

The pneumatic hydromanipulator illustrated in Fig. 26, consists of a hinged mechanism, on one end of which is arranged a pump-dispensing apparatus and on the other, interchangeable pneumatic holders.

The hinged tubular mechanism is analogous to the mechanisms of the pneumatic and hydromanipulators.

The hose-extension roller mechanism has five spindles for simultaneous movement either of the two outer tubes bearing a large double-cavity holder or of three average tubes bearing one of the changeable triple-cavity holders.

The pneumatic holders are moved vertically in the working space by rotation of the leading rollers of the hose-extension mechanism with the handles.

The distributing tap has in its body eight channels for the leads and two coaxial barrels with L-shaped channels, i.e., like a double changeover tap of a known type. Rotation of one of the barrels of the tap connects in turn the cavities of the pneumatic holder and rotation of the other, the dispensers of the pump-dispensing mechanism. When necessary, the system is connected in the same way to the atmosphere or to auxiliary vessels with working media.

The pump-dispensing mechanism consists of two syringe-dispensers with capacities of 20 and 1 ml and one dispenser with a rubber bulb with a capacity of 120-150 ml, which is used in controlling pneumatic holders and

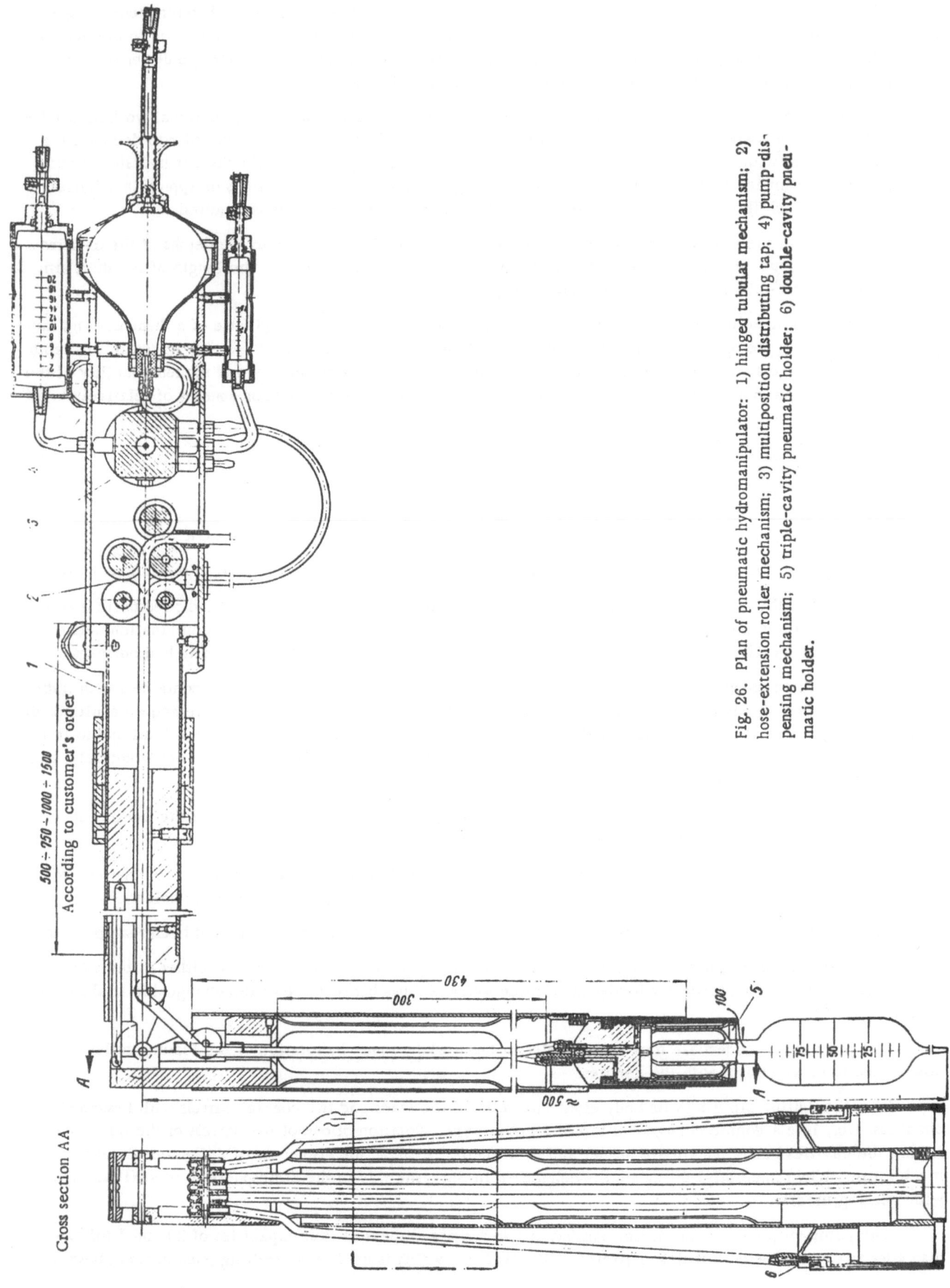

Fig. 26. Plan of pneumatic hydromanipulator: 1) hinged tubular mechanism; 2) hose-extension roller mechanism; 3) multiposition distributing tap; 4) pump-dispensing mechanism; 5) triple-cavity pneumatic holder; 6) double-cavity pneumatic holder.

54

measuring large portions of working media. Syringe-dispensing devices, designed for measuring with interchangeable pipettes (fitted into a triple-cavity pneumatic holder), are used for smaller portions. The volumes are read off either from the scales of the dispensers or those of the pipettes themselves.

Triple-cavity holders (Fig. 26) have inner, outer and central pneumatic working cavities, which are isolated from each other. The inner cavity is used for gripping various articles in the central cavity of the pneumatic holder (externally) and the outer cavity for gripping hollow articles (internally). The feed channel of the central cavity introduces and removes air from the pipettes gripped by the pneumatic holder internally, externally or a combination of the two. By means of the dispensers or other means, the same channel is used to introduce and remove gases and liquids from the outside. The double-cavity holder has an open central cavity and is used for work with articles of large dimensions.

In assembling in position, the manipulator is set in a guide collar, which allows it to be rotated freely around its axis, moved along the axis and fixed rigidly in any position.

Chapter VII

GEAR MECHANISMS FOR MANIPULATING DEVICES

Two groups of simple, versatile and promising manipulating mechanisms have been developed for performing rotary-linear movements in the operation of various devices such as remote grabs, lifting-turning tables, holders, load-handling gear, etc.

The first group is represented by a combined gear-screw mechanism intended for a separate, i.e., circular or linear, remote three-dimensional movement of a driven link and the controlled device attached to it, and a worm-screw drive intended for separate as well as simultaneous circular-linear (i.e., helical) remote three-dimensional movement of the driven link.

The driven link of the first group of mechanisms consists of a screw shaft with a long keyway. Driven through a reduction-reversing gear placed coaxially with it, the shaft rotates about, and moves in the direction of its axis.

The second group of mechanisms, referred to as rack or two-profile gear mechanisms, comprises seven kindred mechanisms.

The distinctive feature of the second group of mechanisms is the use of various types of racks as the driven or the driving link, or as both.

These racks are actually combined gear-racks which can rotate about their axes, or move in the direction of these axes individually or simultaneously; these movements are made possible by the use of two-profile teeth. It is because of this design that the mechanisms belonging to this group are called two-profile or rack mechanisms.

With regard to the arrangement of the teeth on the rack, the mechanisms can be divided into straight-tooth type with two-profile teeth arranged in straight rows parallel to the axis, and cross-tooth or staggered-tooth type with two-profile teeth arranged in checker formation.

The tooth-carrying elements of the racks can be solid or sectional and are designed for power transmission at relatively low speeds.

The mechanisms can consist of two or three links.

The three-link mechanisms are subdivided into: a) circular-rack mechanisms with one straight-tooth rack working with a circular rack and a spur gear (pinion); b) flat-rack mechanisms which are employed for the transmission of larger forces; in these mechanisms a flat rack replaces the circular type; c) circular rack-screw mechanisms, with a screw-threaded member at the end of the circular rack which makes possible an inching rotation of the driven element and prevents undesired rotation; d) worm mechanisms in which a worm is used in place of a circular rack, to make possible an unlimited rotation and self-braking of the racks; e) combined rack mechanisms with a combined rack consisting of an outer, hollow, straight-tooth, flat rack carrying a coaxial gear inside.

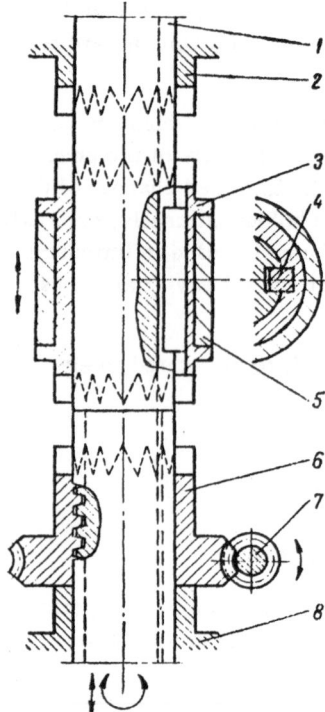

Fig. 27. Gear-screw mechanism for lifting and turning: 1) screw shaft; 2) crown sleeve; 3) claw coupling; 4) sliding key; 5) operating ring; 6) worm wheel; 7) worm; 8) guide sleeve.

The two-link mechanisms can be of staggered-tooth design (with two identical staggered-tooth racks working together) or of combined type (consisting of two different racks — one straight-tooth and the other staggered-tooth).

All mechanisms, with the exception of the self-braking type, can work as reversing mechanisms, i.e., if necessary their driving links can be made into driven links, and vice versa.

All racks with two-profile involute teeth (with the exception of the staggered-tooth type) can be made by any known method, including milling with ordinary disc gear cutters.

1. Gear-Screw Mechanism for Lifting and Turning Operations

This mechanism is intended for performing circular and linear motions, such as the operation of lifting and turning tables, holders, and load-handling devices used in handling, from a distance, radioactive materials in protective chambers.

The driven link of the mechanism is a screw shaft with a longitudinal keyway, while the driving link is a crown worm wheel with an internal thread which is in constant mesh with the shaft. In addition, the mechanism is provided with a fixing link in the form of a stationary crown sleeve and a double-claw coupling which acts as a stopping and reversing link. The coupling is provided with a key which slides in the keyway of the shaft; it is mounted between the faces of the teeth of the worm wheel and of the shaft-guide sleeve.

All links are solids of rotation and have a common geometrical axis.

Figure 27 shows a section of the mechanism through its main axis.
In order to give a clear picture of the operation of the mechanisms, this and the following sections provide detailed descriptions of their operation.

The shaft rotates about its axis, and moves longitudinally in the direction of this axis, in guide sleeves.

The contact faces of the sleeve, coupling, and wheel have meshing rectangular teeth (claws). A reversing motor rotates, through a worm, the worm wheel which rotates the shaft by means of the claw coupling. If the shaft has to transmit upward forces, the worm wheel must be held in position in such a manner as to make its lifting impossible.

In order to give the shaft a linear motion, the coupling is connected to the stationary toothed guide sleeve which prevents the shaft from rotating; since the worm wheel and the shaft have the same thread, the rotating worm wheel will move the shaft in a longitudinal direction.

In the drawing, the double claw coupling is shown in its central position.

The number of teeth on the meshing end faces of the sleeve, coupling, and worm wheel determines the number of positions in which the shaft can move upward and downward and rotate.

The number of teeth of the worm wheel must be the same as the number of teeth on the end faces of the coupling, or a multiple of it; it is desirable that it corresponds to a whole number of degrees of the angle through which a complete turn of the worm rotates the shaft. This makes easier the setting of the shaft by hand into its initial position, and makes the reading of its inching rotation more comfortable.

2. Worm-Screw Mechanism with Double Drive

The worm-screw mechanism with double drive is intended for the circular-linear operation (separately or simultaneously and at different speeds) of various controlled devices, and in particular of remote-controlled lifting and turning equipment.

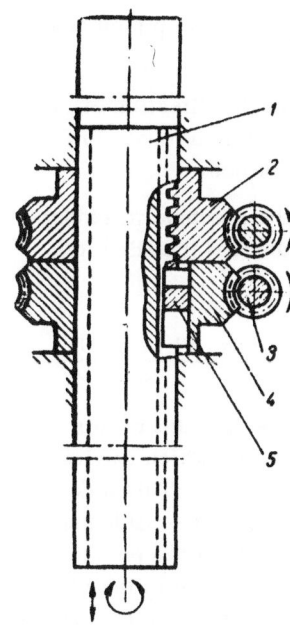

Fig. 28. Worm-screw mechanism with double drive: 1) screw shaft; 2) gear-nut; 3) worm; 4) gear-sleeve; 5) sliding key.

The mechanism being described consists of two driving elements represented by two worm wheels arranged coaxially with the driven screw shaft. The first wheel has an internal thread which is of the same type and diameter as the shaft thread, while the second wheel has a smooth bore corresponding to the outer diameter of the shaft and has a key sliding in a longitudinal keyway.

Figure 28 shows the section of the mechanism through its main axis.

The worms are driven by two identical high-speed reversing motors. The inching motion of the shaft, and especially the fine rotary motion, can also be performed by hand.

The mechanism operates as follows: the rotation of the worm wheel-nut produces a linear motion of the shaft at normal speed, i.e., each revolution of the wheel advances the shaft by the length of the pitch. The rotation of the worm wheel-sleeve produces a helical motion of the shaft with the same speed; if both wheels rotate in the same direction and with the same speeds, then the shaft performs a linear motion; if both wheels rotate with equal speeds but in opposite directions, then the shaft moves helically at double speed, i.e., at the rate of two lengths of pitch per revolution of the wheels.

Other types of shaft motion can be produced by the operator by combining the operation of both motors in different ways. Easy control of the electric motors enables the motions to be performed quickly and independently; in this respect, commutator motors are more flexible than the asynchronous type. By various measures the range of speeds and types of motion can be extended indefinitely. These measures include the use of multispeed motors, gear boxes, etc.

It is easy to arrange the remote control of electric motors and especially the control of the fractional-horsepower type by means of readily movable switches. This eliminates any physical effort usually involved in the control of motors on the part of the operator and renders the coordination of his movements simple and easy. The advantages of this device include the small overall height of the reduction-reversing mechanism, the possibility of push-button control and of using to the full the power of the electric motors.

3. Mechanism with Combined and Two-Profile Racks

This versatile mechanism has one driving and one combined driven link consisting of meshing racks with crossing axes, and is intended for rotary, linear and helical operation of the controlled element connected to the driven link of the mechanism.

The mechanism (Fig. 29) consists of a driving circular rack, with longitudinal and circular crossing teeth constantly meshing with the combined driven rack. The driven rack consists of a spur pinion, with single-profile teeth, which is mounted inside a hollow sleeve-rack (tube). Part of the external tube has a C-shaped cross section. The profile and pitch of the straight teeth cut on the partially open side of the sleeve correspond to those of the driving rack and the internal pinion.

Such a combination is employed to ensure a positive simultaneous, as well as independent, circular and linear motion of the driven link.

The driving rack is in continuous mesh with the pinion and the hollow rack, which both form a kind of combined rack. The hollow rack can perform only a linear motion in the direction of its axis. The pinion can freely rotate inside the rack but can move in a linear direction only together with this rack.

Both racks (the driving and the combined) are mounted in a body link (Fig. 29).

When the driving rack rotates in one or the other direction, the driven combined rack moves upward or downward, while the pinion with its straight teeth remains stationary. When the driving rack moves backward or forward the pinion revolves, while the sleeve-rack remains stationary.

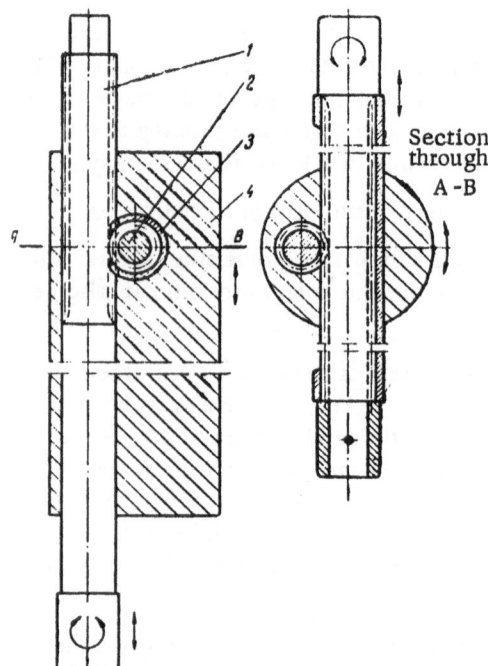

Section
through
A -B

Fig. 29. Combined-rack two-profile
mechanism: 1) driving rack; 2) pinion;
3) hollow sleeve-rack; 4) body link.

During helical motion of the driving rack, the pinion and
its rack perform a linear motion.

The body link can revolve or move in the direction of its
axis, thus ensuring the possibility of a three-dimensional motion
of the controlled elements attached to the pinion.

The versatility and compactness of the mechanism render
it especially valuable for use in handling devices intended for
various purposes.

4. Two-Profile Gear Mechanism for Three-Dimensional Manipulating

The distinctive feature common to mechanisms of this
group is the use of a multifunctional driven link which carries
the controlled elements on one or both of its ends. This link is
in the form of a long combined pinion-rack with four-sided two-
profile teeth. In the radial section the teeth have the profile of
spur gears (for example, an involute contour), while in the axial
section their profile is that of a rack (for example, trapezoidal).
Such a design of the pinion teeth makes possible a rotary, linear,
or combined (i.e., helical) motion of the driven link of the
mechanism.

For various service conditions and kinds of drive, four types
of rack mechanisms were developed: circular rack, circular rack-
screw, flat rack, and worm mechanisms.

The circular-rack mechanism is the most widely and universally used device. It is shown in two sections
in Fig. 30.

The purpose of the body link is to house and to move the other three links in circular and linear directions.

The pinion rotating about its axis moves the rack in an axial direction. The linear motions of the circular
rack rotate the operating rack in either direction.

The simultaneous motion of both driving links of the mechanism (pinion and circular rack) produces a
helical motion of the driven rack.

The circular rack-screw mechanism (Fig. 30, II) is also intended for performing three-dimensional handling
operations similar to those described above. In this mechanism one end of the circular driving rack is threaded;
the purpose of this screw thread is to provide for inching and self-braking of the driven rack. This member,
which passes through a threaded bore in the mechanism body, provides a safeguard against an accidental movement of the circular rack in an axial direction and the rotation of the driven rack.

The flat-rack mechanism is obtained by replacing the circular-rack in the mechanism (Fig. 30) by a type
with straight trapezoidal teeth. Such a system is capable of transmitting greater loads owing to the increased
number and stronger cross section of the circular-rack and driven-rack teeth which are in mesh at any given moment.

The worm mechanism is obtained by replacing the circular rack in the mechanism (Fig. 30, I) by a worm
which meshes with the driven rack; it is arranged according to its lead angle and prevented from axial displacement by suitable means. The use of a worm ensures an unlimited range of fine adjustment of the driven rack,
as well as its self-braking.

5. Staggered-Tooth Rack Mechanism

These two-profile gear mechanisms with normal and improved external teeth are designed also for circular, linear and helical motions. The racks have the normal or twice the normal number of teeth continuously
in mesh.

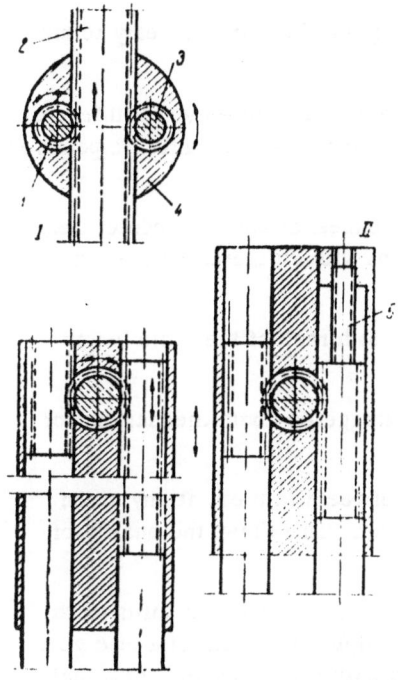

Fig. 30. Two-profile gear mechanisms for three-dimensional manipulating. I — Circular-rack mechanism: 1) gear; 2) rack; 3) circular rack; 4) body link. II — Circular rack-screw mechanism: 5) screw link.

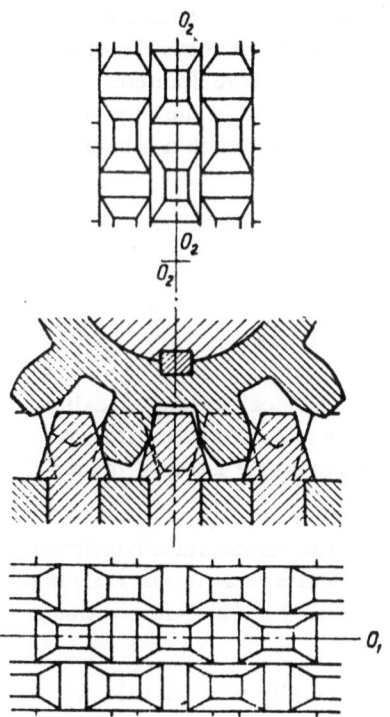

Fig. 31. Sectional staggered-tooth mechanism.

According to the service conditions and the design of the mechanisms they can be classified as follows:

1) by the method of combining the racks into staggered-tooth and combined mechanisms;

2) by the design of toothing into solid and sectional types.

The sectional racks of staggered-tooth mechanisms consist of similar or different (with regard to their shape and dimensions) coaxially joined symmetrical or asymmetrical (with regard to thickness) gear rings, with crown or peripheral teeth.

Sectional staggered-tooth racks are sometimes used in combination with straight-tooth racks in combined mechanisms in order to make them simpler and to reduce their cost.

All types of two-profile racks can be made by any known gear-cutting method, including milling with disc gear cutters.

Figure 31 shows the section through the involute teeth of sectional staggered-tooth racks working together. The section is through the $O_1 - O_1$-axis of one of the racks, and is perpendicular to the O_2-axis of the other rack. The drawings below and above the section show developed parts of the surface of the lower (first) rack and of the lower surface of the second rack, respectively.

The sectional racks are assembled with identically toothed rings arranged coaxially and joined with one another. These rings represent the individual teeth of a circular hollow rack with the teeth of a spur gear cut on its periphery.

The overall thickness of a ring is equal to the nominal thickness of a rack tooth at its root (2.44 x module). The end faces of each ring are counterbored to a diameter which is equal to the spaces between the circular rack teeth, and to a depth which is 0.435 x x module. These counterborings, or recesses, leave in the central part of the rings a thickness of 1.57 x module which is equal to half of the pitch. Instead of one large recess, individual recesses can be provided at each gear tooth, their number must thus be equal to the number of teeth on the ring's circumference, the angular width should be equal to the angular pitch of the teeth and the depth should be 0.435 x module (a combination of both designs can also be used).

In the case of an ordinary involute profile the root flanks of teeth within the pitch circle are plane and the width of a tooth space on the pitch circle is nominally equal to the tooth thickness. This fact, as well as good finish of the ring end faces, makes possible their assembly into a hollow rack. In a rack the rings are assembled in such a manner that their two-profile teeth are arranged together in a staggered formation; the end faces of the teeth of two adjoining rings fit into the tooth spaces of one another. The rack is prevented from displacement along the axis and rotation on the shaft by one of the conventional methods (for example, by tightening with a nut, or the use of a key).

The toothed rings of the described sectional racks can have a different shape and be assembled in a different way.

Crown rings, asymmetrical, but similar in shape and dimensions, with an overall thickness of 2.44 x module and a recess of 0.87 x module having one flat face, are easy to produce by individual machining.

Crown and toothed rings (i.e., flat on both sides) placed alternately and having the same overall thickness of 2.44 x module but with different thicknesses of the central part (0.7 x module and 2.44 x module, respectively), i.e., with teeth of different strength, are also made by machining.

The symmetrical toothed rings similar in shape and dimensions, with a thickness of 1.57 x module, i.e., with rack teeth slightly reduced in thickness, which are intended for transmissions with an increased clearance between the teeth working together, are the easiest to make.

A staggered-tooth rack with the number of teeth twice that of a straight-tooth rack of the same length can be made from the latter.

A staggered-tooth rack can work with another staggered-tooth rack (as a staggered-tooth mechanism) or with a straight-tooth rack (as a combined mechanism).

If a straight-tooth and a staggered-tooth rack work together, the number of teeth in mesh at any given moment is half the number meshing when two staggered-tooth racks work together. This shows the quality of gearing for both types of mechanisms.

A solid rack with teeth arranged in regular rows (i.e., a straight-tooth rack) can be made of solid-tooth circular rack by milling additional tooth spaces in it. This can be done, for example, by means of a disc gear cutter. A solid-tooth circular rack is made on a lathe using a disc gear milling cutter as a circular form tool.

When racks are made in large quantities it is economical to use more efficient manufacturing methods; the casting and hot rolling of solid racks, and hot stamping of sectional-rack rings, the use of special rolled sections as raw material, etc. In casting or hot rolling staggered-tooth solid racks, sectional racks can be used as models or rollers.

In combined mechanisms the staggered-tooth rack is used in the shorter (driving) link, since the driven link is usually required to make only one or two or, at the most, a few revolutions, while its linear motion extends over the entire length.

If the driving link is placed inside a cylindrical hollow body-member of the mechanism with the driven link arranged perpendicular to them both, the latter can move parallel to its own axis in the direction of and about the axis of the body link (provided that this member moves in a guide bore of the same diameter). If the body member is mounted on a sphere, the driven link of the mechanism can also move within the required solid angle, with the apex in the center of the sphere. Utilizing the described advantages of the rack mechanism it can be widely used in handling and manipulating devices intended for various purposes.

CHAPTER VIII

NONWASTEFUL METHODS OF AMPOULE INSPECTION

Methods are described and devices designed for individual or group and total or local testing of sealing, mechanical strength and thermal stability of sealed glass ampoules with radioactive and other toxic liquids. The ampoules are checked remotely in shielded, exhausted cupboards without irrecoverable losses of the ampoule contents.

For a normal set of checks, the ampoules are subjected to artificial pressure and temperature conditions, imitating normal or the worst possible transport and working conditions for the ampoules so as to test their mechanical and thermal stability; tests are also made for permeability to gases and liquids by visual observation and automatic self-registration of defective ampoules.

The necessity for such a set of checks on ampoules, including solid ones, especially with volatile, viscous and emanating liquids, is one of the basic requirements of radiation safety.

With rigid control, the working cycle of normal checks is repeated several times. A simplified control differs in that the working cycle is reduced.

The normal set of ampoule checks begins directly after sealing of the filled ampoules, with a preliminary inspection before and after cooling.

The first check of the working cycle must be cooling of the ampoule in an inverted position to a temperature below that of the surrounding medium. This operation is usually done by the so-called dry method, in a cooling chamber with a stream of carbon dioxide from a bulb with liquid and solid carbon dioxide (dry ice), to determine the thermal stability of the material of the ampoules at a low temperature, down to the required limit. The cooling may be total or local, especially in the sealing zone. The stability of the ampoules to excess external pressure is thus tested by creating a certain reduction in pressure in them by cooling.

In defective ampoules, pressure equilibrium is attained by the leakage of air through the walls (through openings, cracks, micropores in the glass, etc.). The passage of air into the ampoules may be observed by the appearance of air bubbles in the liquid. The second control operation of testing the resistance of the glass to increased surface temperature and heating through is carried out by heating the ampoule by a dry method, e.g., infrared lamps, hot air, etc. Defective ampoules leak.

The check may be visual or autochromatographic — by the color of spots from liquid flowing from the ampoules onto filter or indicator paper. The latter is used to give more noticeable spots. The paper may be placed against the ampoules or at a distance.

In checking ampoules with low-boiling liquids, it is better to have the paper in immediate contact with the ampoule. This gives the most reliable check on small drips and if the ampoule bursts, the contents are caught for subsequent use. For this purpose, a clean cell, made from a material inert to the ampoule contents, should be placed below the ampoule. For ampoules with acid or alkaline contents, normal litmus paper may be used and in other cases, paper with appropriate neutral indicators.

The operations may be observed and controlled through shielding observation glasses, where necessary using appropriate optical devices, e.g., normal and specially adapted binoculars. The temperature may be controlled directly by observing misting and icing of the ampoules, the use of thermal indicators, electrothermometers, etc.

Auxiliary thermal and pressure control may be provided by the working conditions, i.e., the consumption of the form of energy used and its time of action.

The working conditions are established according to the given control parameter and maintained with appropriate measuring and controlling equipment. The reserve factor is determined from experimental pressure tests with blow-out ampoules.

From this it is clear that, in principle, the process of a normal set of checks on ampoules with liquids consists of two operations — cooling and heating.

The simplified check on the ampoules consists of one operation — heating by the method described above. Naturally this test is less reliable since thermal tests of the ampoule by cooling and pressure tests by increasing the external pressure are not carried out. In addition, the internal pressure of sealed ampoules is always slightly low, due to the expansion of air in them during sealing and its rarefaction on cooling. Therefore, in the heating check, the internal pressure in well-sealed ampoules may not reach the value required for the pressure test at the given temperature. In defective ampoules, in the simplified check, the pressure may be insufficient to cause leakages, especially through microholes. This may be partially compensated by allowing the ampoules to stand for some time, to reach pressure equilibrium, or combining this with inspection for the infiltration of air as described above (with ampoules upside down).

Semi-rigid control of ampoules combines the simplified and the normal methods in that order, i.e., it consists of three operations and further explanation is not required.

Rigid control of ampoules differs from the other methods in that the working cycles of normal control are repeated the required number of times, which guarantees the best test of the ampoule quality in comparison with the other methods described. The number of the operations described above must always be sufficient for a fully reliable check.

Checking of ampoules with dry contents includes testing for impermeability by forcing appropriate liquid indicators into defective ampoules under pressure and also heating with subsequent cooling.

When the contents of defective ampoules have to be kept in a fit state for subsequent use, the indicators must always be neutral or harmless.

One of these methods is the normal set of pressure and temperature checks for ampoules. The simplest and most convenient neutral indicator, which is quite acceptable in many cases, is distilled water (a wetting indicator) as it readily penetrates microholes and produces noticeable agglomeration of the ampoule contents, especially with powdery or finely granular products. Thus it is only necessary to partly wet the substance with the indicator and, best of all, to have intimate contact between the indicator liquid and the place where the ampoule was sealed.

In testing the ampoule, the neck should be placed downward as it is the most probable place for defects.

The other simplified method of pressure and temperature checking differs from the first in that it involves only the second operation — cooling the ampoule, immersed to the neck in indicator liquid.

The third method, semi-rigid pressure and temperature checking, consists of three operations and is a combination of the simplified and normal methods.

The fourth method, rigid checking, consists of performing several working cycles of the normal check.

CHAPTER IX

SOME DECONTAMINATION METHODS

1. Remote Hydromechanical Cleaning of Contaminated Surfaces

Systematic wet cleaning of actual production rooms and especially rooms used by many is one of the most important health physics measures. In many cases it is desirable and often necessary that the wet cleaning of rooms and equipment is done remotely, i.e., without direct contact between the operator's hand and the surfaces cleaned and the cleaning media.

Wet cleaning of rooms in which radioactive substances are used and processed (laboratories, vivariums for experimental animals, wash and shower rooms, etc.) is of extreme importance. The choice of wet cleaning method inevitably depends not only on the degree of cleanness required and on the health of the staff and accessory personnel, but also on the possibilities of localization of specific contaminants and prevention of their gradual accumulation and spreading.

Methods of wet cleaning of specific contaminated rooms with alkalis, volatile and hot washing materials used at the present time are dangerous to the health of workers and rubber gloves are not a reliable protection of the organism from harmful effects and are inconvenient in use. In addition, the cleaning of large areas (floors of rooms) by hand requires a great and inefficient consumption of labor and time.

As a result of what has been described, a need arose for producing simple, portable and cheap multifunctional hand appliances, convenient for remote hydromechanical cleaning of contaminated surfaces. The use of such a cleaner, in combination with other appliances, ensures cleaning of the room with the minimum expenditure of labor and time.

The appliance described satisfies these requirements and consists (Fig. 32) of two metal or wooden rods, 20-25 mm in diameter and 0.75-1.5 m or more in length. To one end of each rod is attached a strip of cloth, 40-45 cm wide and 60-80 cm long with good water absorbing properties.

If thin cloth is used, it may be first folded several times to give the required thickness and width. The three last folds in the cloth are made concertina-wise so that the four sections of the fold have a cross section similar to the Latin letter W and the width of the longitudinally folded strip is 1 dm.

The strip of cloth in this form is bound to the rods as shown in Fig. 32.

This form of cloth arrangement is used largely in treating large surfaces (floors and walls of rooms) and also in rinsing the cloth.

Fig. 32. Treating a floor.

The cloth band is squeezed out by rotating one of the rods around its longitudinal axis like a solid roller (Fig. 33).

In treating small, high surfaces, such as table-tops, window sills, cupboard walls, etc., the strip of cloth is twisted into a roll on each rod (Fig. 34).

The cloth may be twisted onto one rather than the two rods and, as it becomes contaminated, it may be rewound onto the other rod until one side of the cloth is completely contaminated and then wound back the other way so as to use both sides of the cloth.

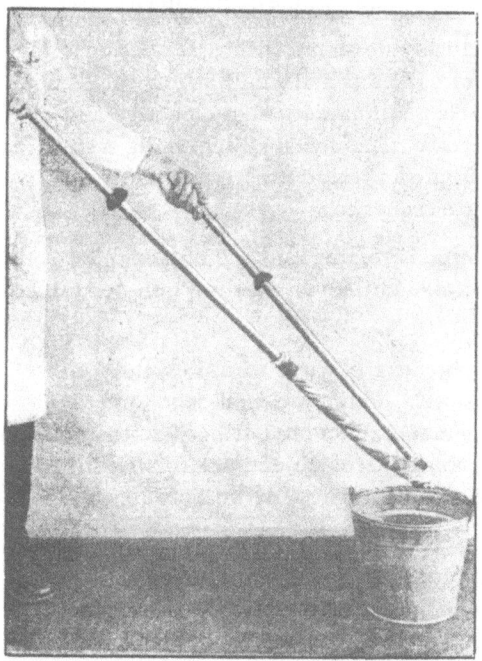

Fig. 33. Wringing out the cloth by twisting it with the rods.

To treat small surfaces (channels, cavities in vessels, etc.), the cloth may be twisted up and then coiled around one of the rods (Fig. 35).

Longer rods are best made from thin-walled stainless steel, dural or bamboo.

With frequent changes of the cloth, the rods are best made in parts, with the lower section, to which the cloth is attached, made of wood in the form of a readily changeable tip, fitting tightly into a metal tube. In the end of a rod is made a diametrical slit, 3-4 mm wide and 50 mm deep, and two holes, 3-4 mm in diameter, perpendicular to the plane of the slit. The cloth is passed through the slit and strong cord or twine through the holes to attach the strip of cloth to the rod tightly.

Sometimes it is necessary to have the rods in contact and fix them in this position. For this purpose, two perpendicular metal pins are set in one of the rods and two pairs of similar holes drilled in the other. The pins of one rod fit into the holes of the other, making their attachment convenient.

To prevent the possibility of washing solutions falling on the hands of the worker (in cleaning window sills, cupboards, etc.), protective flexible traps for liquid draining along the rods are attached to the rods (Fig. 34).

One of the advantages of the appliance is the possibility of removing dirt and excess moisture by simple wringing of the cloth with the help of the rods.

The "self-cleanability" of the appliance makes it suitable for treatment of difficultly accessible places

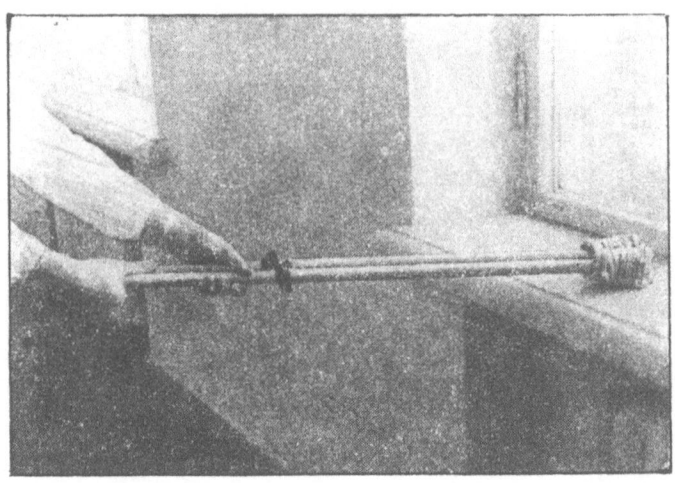

Fig. 34. Treatment of a window sill.

and for cleaning shoe cupboards at the exit from contaminated rooms and at the entrance to rooms where special cleanliness must be maintained.

A worn or contaminated cloth may be removed from the rods by cutting the bindings.

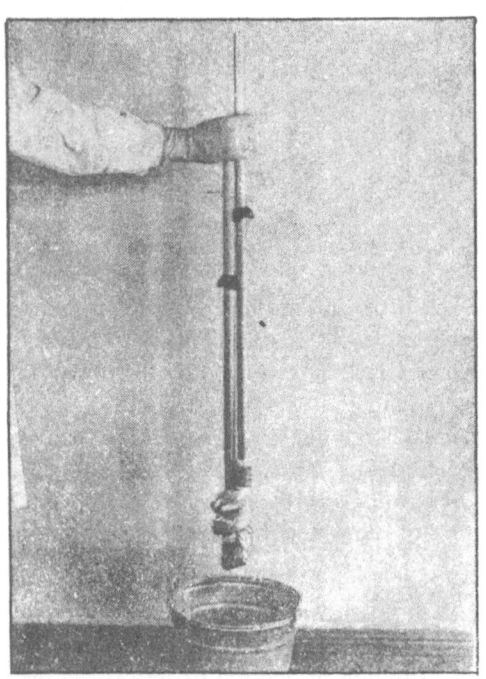

Fig. 35. Coiled form of twisted cloth for treatment of cavities.

Contaminated wooden rods are destroyed together with the cloth, and metal rods may be freed from contamination and used again.

2. Multisolution Stationary Decontaminators

Multisolution stationary decontaminators are designed for deactivating the hands of operators and small articles of laboratory and production use contaminated with radioactive and other highly toxic materials. The decontamination is carried out with a set of washing media (solutions are previously prepared) and in a definite sequence.

The washing materials are prepared and distributed among designated places and collected after use in a centralized fitting.

Decontaminated objects are wiped with previously prepared paper and gauze towels, which are used once, and also with cotton and cotton gauze wads, which are collected in special waste receptacles. Decontaminated objects are finally dried with combined infrared and hot-air dryers, working separately or simultaneously.

The washing solutions from the leads enter separate standard tanks. The number of tanks is determined by the number of decontaminating solutions used and does not usually exceed five. The tanks are fitted with a ball-valve and a self-clearing lifting plug, which, respectively, keep the volume of the liquid in the reservoir constant and dispense a suitable amount of washing solution, when the plug is operated.

In an improved modification of the deactivator, a single multichamber tank is used, which is made in accordance with sanitary requirements applying to equipment used for work with radioactive materials in laboratory and industrial conditions. In simplified deactivators, tanks of the wash-stand type with lifting conical plugs may be used as reservoirs for the washing solutions. The inflow of washing solutions is controlled with foot-operated valves.

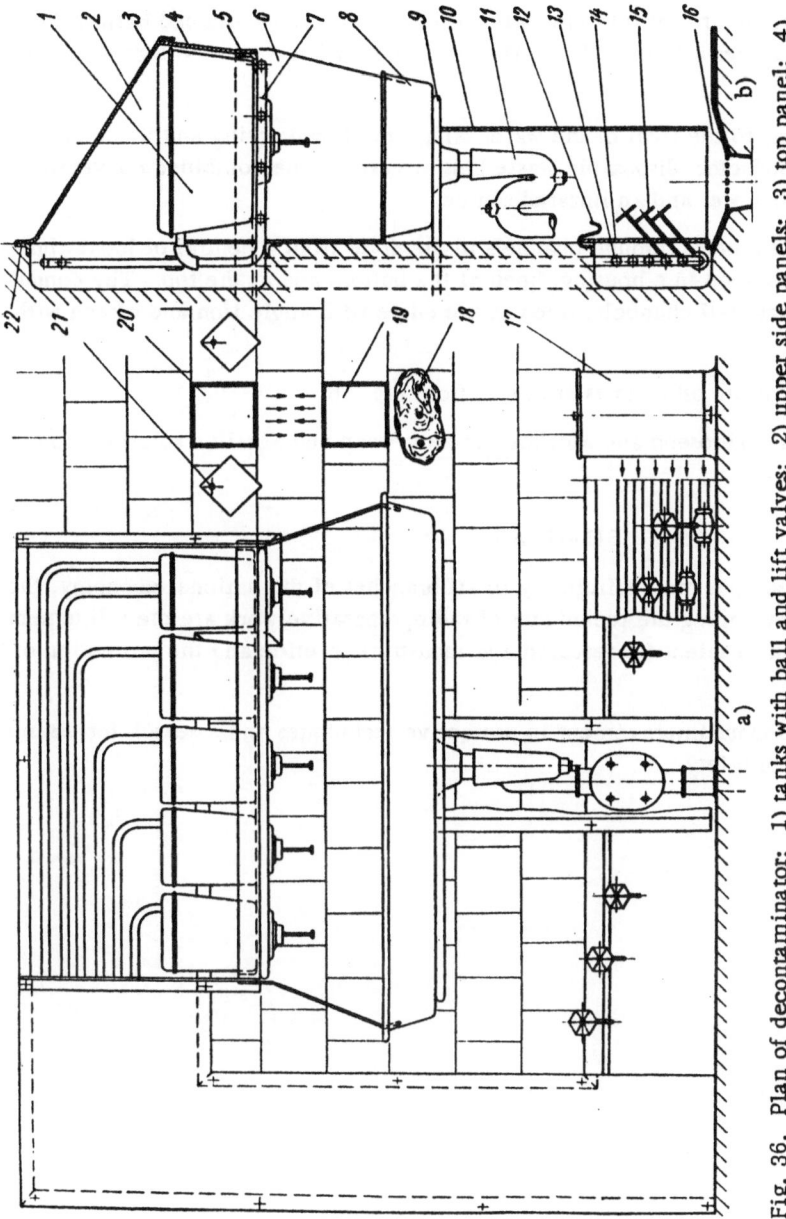

Fig. 36. Plan of decontaminator: 1) tanks with ball and lift valves; 2) upper side panels; 3) top panel; 4) front panel; 5) angled frame for tanks; 6) anti-splash cover; 7) heating register; 8) trough-shaped basin; 9) tubular frame for basin; 10) syphon cover; 11) syphon with inspection fitting; 12) under-collector trough panel; 13) water and solution tanks; 14) collector cover; 15) foot-controlled valve; 16) drainage gutter with trap; 17) pedal waste receptacle; 18) trident for cotton; 19) hot-air dryer; 20) infrared dryer; 21) holders for paper and gauze towels; 22) facing angle irons.

65

Figure 36 (<u>a</u> – front view, <u>b</u> – side view) illustrates one form of decontaminator.

The tanks may be partly inset into a wall recess or arranged parallel to rather than perpendicular to the basin. In the extreme right tank are two ball-valves feeding in hot and cold water. The hot water is first passed through a heat register, which is fixed underneath the frame and passes underneath all the tanks. The amount of hot and cold water passing into the right tank is controlled with valves. The knobs of the valves are rotated with the foot by means of a form of crank. The other tanks are operated similarly. At the side, the top and front are closed with covers, which are readily changed. Their surfaces, like the surfaces of all the parts, must be smooth and readily washed. The upper surface is inclined so that no form of contaminated article can be placed on it.

The ends of the angle-iron frame are set into the wall. The standard trough-shaped basin is held with a tubular frame. The drainage gutter goes all along the wall of the room and serves to collect water from washing the walls and floor.

Near to the washing facilities, to the right of the water tank, are placed drying and waste collection facilities. In the absence of a waste conductor, disposable waste bins are used. The combined convection-radiation dryers consist of a so-called hot-air towel and an infrared source.

A stream of hot, dry air flows upward through channels in the dryer from a central or local apparatus, e.g., a calorifier. The infrared lamp is placed in a housing, open at the bottom and at the top. The conduits for the water and solution leads are made as wall channels, lined at the edges with angle iron and closed with readily removable covers attached to this.

The design and construction of the other parts are clear from Fig. 36.

The washing solutions used are kerosene and alcohol <u>kontakty</u>,*soap and alkaline solutions and also other washing media.

SUMMARY

In order to use radioactive isotopes most efficiently in all branches of the national economy, the main problems of the present stage of developing the techniques of radiopreparative work are the refinement and adoption of new techniques and methods in scientific research and industrial practice and the protection of health from specific harm.

This creates a need for continuous improvement in protective techniques and a search for essentially new working methods to ensure complete safety.

---

*Sulfonic acid derivatives.

66